BIOTECHNOLOGY
INTELLIGENCE
UNIT 7

Synthetic DNA Delivery Systems

Dan Luo
Cornell University
Ithaca, New York

and

W. Mark Saltzman
Yale University
New Haven, Connecticut

LANDES BIOSCIENCE
GEORGETOWN, TEXAS
U.S.A.

EUREKAH.COM
AUSTIN, TEXAS
U.S.A.

SYNTHETIC DNA DELIVERY SYSTEMS

Biotechnology Intelligence Unit 7

Landes Bioscience / Eurekah.com
and
Kluwer Academic / Plenum Publishers

Kluwer Academic / Plenum Publishers, 233 Spring Street, New York, New York, U.S.A. 10013
http://www.wkap.nl/
Please address all inquiries to the Publishers:
Eurekah.com / Landes Bioscience, 810 South Church Street, Georgetown, Texas, U.S.A. 78626
Phone: 512/ 863 7762; FAX: 512/ 863 0081; www.Eurekah.com; www.Landesbioscience.com

ISBN: 0-306-47701-7

Library of Congress Cataloging-in-Publication Data

Synthetic DNA delivery systems / edited by Dan Luo and W. Mark Saltzman.
p. cm.—(Biotechnology intelligence unit ; v. 7)
Includes bibliographical references and index.
ISBN 0-306-47701-7
1. Genetic engineering. 2. Transfection. 3. Drug delivery systems.
I. Luo, Dan. II. Saltzman, W. Mark. III. Series.

QH443.S96 2003
660.6′5—dc21 2002043446

CONTENTS

EDITORS

Dan Luo
Department of Biological and Environmental Engineering
Cornell University
Ithaca, New York

and

W. Mark Saltzman
Departments of Chemical Engineering and Biomedical Engineering
Yale University
New Haven, Connecticut

CONTRIBUTORS

Kristian Berg
Department of Biophysics
The Norwegian Radium Hospital
Montebello
Oslo N-0310
Norway
Chapter 8

Gennaro Ciliberto
Department of Molecular and Cellular Biology
Istituto di Ricerche di Biologia Molecolare
I.R.B.M.-P. Angeletti
Via Pontina Km. 30,600
Pomezia (Roma) 00040
Italy
Chapter 6

Matt Cotten
Institute for Molecular Pathology (IMP)
Dr. Bohr-Gasse 7
Vienna A-1030
Austria
Chapter 5

Dominique Coulaud
Laboratoire de Microscopie Moléculaire et Cellulaire
Institut Gustave Roussy
UMR 8532 CNRS
rue Camille Desmoulins
Villejuif cedex
F-94805
France
Chapter 3

Etienne Delain
Laboratoire de Microscopie Moléculaire et Cellulaire
Institut Gustave Roussy
UMR 8532 CNRS
rue Camille Desmoulins
Villejuif cedex
F-94805
France
Chapter 3

Sandra S. Diebold
Imperial Cancer Research Fund (ICRF)
44 Lincoln's Inn Fields
London WC2 3PX
United Kingdom
Chapter 5

Birgit Ø. Engesæter
Department of Biophysics
The Norwegian Radium Hospital
Montebello
Oslo N-0310
Norway
Chapter 8

Elena Fattori
Department of Molecular and Cellular Biology
Istituto di Ricerche di Biologia Molecolare
I.R.B.M.-P. Angeletti
Via Pontina Km. 30,600
Pomezia (Roma) 00040
Italy
Chapter 6

James F. Greenleaf
Department of Physiology and Biophysics
Mayo Foundation
Rochester, MN
Chapter 9

Marit Hellum
Department of Biophysics
The Norwegian Radium Hospital
Montebello
Oslo N-0310
Norway
Chapter 8

Anders Høgset
Department of Biophysics
The Norwegian Radium Hospital
Montebello
Oslo N-0310
Norway
Chapter 8

Ban-An Khaw
Department of Pharmaceutical Sciences
Bouve College of Health Sciences
Northeastern University
Boston MA 02116, USA
Chapter 4

Eric LeCam
Laboratoire de Microscopie Moléculaire et Cellulaire
Institut Gustave Roussy
UMR 8532 CNRS
rue Camille Desmoulins
Villejuif cedex
F-94805
France
Chapter 3

Gary Lee
Chemical Engineering and the Wills Neuroscience Institute
University of California at Berkeley
Berkeley CA
94720-1462
USA
Chapter 2

Patrick Midoux
Centre de Biophysique Moléculaire
UPR 4301 CNRS
rue Charles Sadron
Orléans Cedex 02
F-45071
France
Chapter 3

Douglas L. Miller
Department of Radiology
University of Michigan
Ann Arbor, MI
Chapter 9

Nicola La Monica
Department of Molecular and Cellular Biology
Istituto di Ricerche di Biologia Molecolare
I.R.B.M.-P. Angeletti
Via Pontina Km. 30,600
Pomezia (Roma) 00040, Italy
Chapter 6

Manfred Ogris
Pharmaceutical Biotechnology
Center for Pharmaceutical Research
Ludwig-Maximilians-University Munich
Butenandstr. 5-13
Munich D-81377
Germany
Chapter 7

Vibeke M. Olsen
Department of Biophysics
The Norwegian Radium Hospital
Montebello
Oslo N-0310
Norway
Chapter 8

Chantal Pichon
Centre de Biophysique Moléculaire
UPR 4301 CNRS
rue Charles Sadron
Orléans Cedex 02
F-45071
France
Chapter 3

Sorin V. Pislaru
Department of Physiology and Biophysics
Mayo Foundation
Rochester, MN
Chapter 9

Bruno Pitard
Faculté de Médecine
Institut National de la Santé et de la Recherche Médicale
Unité 533
1 rue Gaston veil
Nantes cedex 1 44035
France
bruno.pitard@nantes.inserm.fr
Chapter 1

Christian Plank
Institute for Experimental Oncology and Therapy Research
Technical University Munich
Ismaninger Str. 22
Munich D-81675
Germany
Chapter 5

Lina Prasmickaite
Department of Biophysics
The Norwegian Radium Hospital
Montebello
Oslo N-0310
Norway
Chapter 8

David Schaffer
Chemical Engineering and the Wills Neuroscience Institute
University of California at Berkeley
Berkeley CA 94720-1463, USA
Chapter 2

Torunn E. Tjelle
Department of Biophysics
The Norwegian Radium Hospital
Montebello
Oslo N-0310
Norway
Chapter 8

Carlo Toniatti
Department of Molecular and Cellular Biology
Istituto di Ricerche di Biologia Molecolare
I.R.B.M.-P. Angeletti
Via Pontina Km. 30,600
Pomezia (Roma) 00040
Italy
Chapter 6

Vladimir P. Torchilin
Department of Pharmaceutical Sciences
Bouve College of Health Sciences
Northeastern University
Mugar Building, Room 312
360 Huntington Avenue
Boston MA 02115, USA
vtorchil@lynx.neu.edu
Chapter 4

Ernst Wagner
Pharmacy
Center for Drug Research
Ludwig-Maximilians-University Munich
Butenandstr. 5-13, Building D, Room 3.030
Munich D-81377
Germany
ernst.wagner@cup.uni- muenchen.de
Chapters 5 and 7

Volkmar Weissig
Department of Pharmaceutical Sciences
Bouve College of Health Sciences
Northeastern University
Boston MA 02117, USA
Chapter 4

Carl J. Wheeler
Vical Inc.
9373 Towne Centre Drive
San Diego CA 92121
USA
Chapter 8

Martin Zenke
Max-Delbrück-Centre for Molecular Medicine (MDC)
Robert-Rössle-Str. 10
Berlin D-13125
Germany
Chapter 5

Biotechnology Intelligence Unit

Other Volumes in this Series

Erythrocyte Engineering for Drug Delivery and Targeting
Edited by Mauro Magnani

Synthetic Polymers for Biotechnology and Medicine
Edited by Ruth Freitag

Adaptive Systems in Drug Design
By Gisbert Schneider and Sung-Sau So

Tracking Genetically-Engineered Microorganisms
Edited by Janet K. Jansson, J.D. van Elsas and Mark Bailey

Recombinant Protease Inhibitors in Plants
Edited by Dominique Michaud

Molecular Responses to Cold, Drought, Heat and Salt Stress in Higher Plants
Edited by Kazuo Shinozaki and Kazuko Yamaguchi-Shinozaki

PREFACE

Significance and Motivation for This Book

The introduction of exogenous DNA into cells (DNA delivery) is among the most important techniques in biomedical sciences. For example, DNA delivery is a powerful and popular tool in elucidating gene structure, regulation, and function. DNA delivery is also destined to become an important field in clinical medicine. With the human genomes sequenced, the bottleneck of gene therapy as well as other DNA-based medicine has shifted from gene cloning to gene delivery. Among the myriad DNA delivery approaches developed so far, viral methods are well known and can be extremely efficient (viral vectors were used in the first human gene therapy test), but the safety (including the toxicity and immunogenecity) of viral vectors is unknown. As a result, nonviral approaches to DNA delivery (i.e. transfection) are becoming particularly popular; indeed, a recent search of key word "transfection(s)" using the National Center for Biotechnology Information's (NCBI; Rockville, MD) MEDLINE database (www.ncbi.nlm.nih.gov/PubMed/ on Jan. 20, 2003) yielded more than 80,317 papers.

The search for alternate nonviral synthetic approaches to DNA delivery has been in progress since the late 1950s and now includes transfection-enhancing chemicals, such as lipids and polymers, as well as mechanical and electrical methods for enhancement. Unfortunately—as we describe in more detail in a recent review[1]—synthetic DNA delivery systems are usually inefficient and toxic (Fig. 1). Therefore, a thorough discussion of barriers along the DNA delivery pathway and a detailed review of what has been achieved, along with a new look at other as-yet-not-so-popular methods, will certainly help us understand the fundamental mechanisms of DNA delivery, and thus point out new directions in devising novel synthetic DNA delivery systems that are safer and more efficient.

In this volume, we have assembled a variety of reviews and perspectives on the current status and future prospects of DNA delivery. The use of lipid-based transfection systems is well-studied and the subject of several recent books.[2] We decided to focus our attention on other synthetic techniques, in which the fundamentals have not been extensively explored.

The barriers along the DNA delivery pathway lie on three major levels: 1) extracellular (DNA condensation and complexation); 2) intracellular (endocytosis); and 3) nuclear (nuclear targeting and nuclear entry). On the extracellular level, negatively charged DNA molecules must first be condensed (usually via complexing with cationic transfection reagents) into particles small enough to be taken up by cells (usually less than 1 μm). On the intracellular level, endocytosis—a multistep process involving binding, internalization, formation of endosomes, fusion with lysosomes, and lysis—is the major route of DNA entry, which determines subsequent DNA release, trafficking, and lifetime within the cytosol. The extremely low pH and enzymes within endosomes and lysosomes pose serious challenges to DNA integrity. On the nuclear level, DNA that has survived the

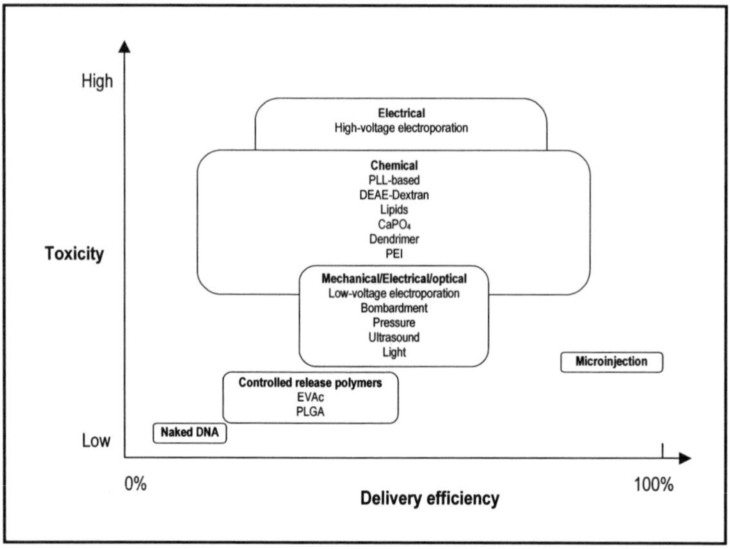

Fig. 1. Comparison of efficiency and toxicity of various nonviral DNA delivery approaches (adapted from the figure of the review (1)).

intracellular path must then dissociate from the condensed complexes, localize the nucleus, and enter through nuclear pores (which are ~10 nm in diameter). DNA may also be recruited into the nucleus during the physical process of cell division.

Only DNA that survives all these barriers has the potential to be functional. As a result, the number of DNA molecules decreases at each step of the journey to the nucleus. Therefore, identifying and overcoming each barrier along the pathways can significantly increase DNA delivery efficiency.

Emphasis of the Book

The effect of delivery systems on these major barriers to DNA delivery is the focus of the first four chapters. In Chapter 1, Pitard reviews the physicochemical properties of self-assembled delivery systems on the basis of cationic lipids and polymers. In Chapter 2, Lee and Shaffer outline the obstacles that must be overcome in designing targeted DNA delivery systems by modification of surface properties of the DNA delivery system. In Chapter 3, Midoux et al. describe the use of histidine-rich peptides and polypeptides as DNA delivery systems; these materials complex DNA and destabilize membranes, thereby overcoming key barriers. In Chapter 4, Torchilin et al. describe gene delivery systems that target intracellular sites such as the cytoskelton or mirochondria.

Equally important as efficient delivery is targeted delivery, in which specific molecular agents are used to enhance the delivery of DNA to a specific target. Targeting is the focus of the three chapters. The first, by Diebold et al., focuses on dendritic cell-targeted gene delivery. The second of these chapters, by Fattori et al., focuses on muscle cell transfection, which is particularly important for DNA vaccination. The third section, by Ogris and Wagner, focuses on tumor-targeted DNA delivery, which is critically important for cancer therapy.

The last two chapters focuses on two DNA delivery systems that have not been extensively explored in the past. In both cases an external signal is used to enhance delivery. The chapter by Hogset et al. describes a technology for utilizing light to direct DNA delivery. The chapter by Miller and Greenleaf uses sound waves as the signaling mechanism. These two new approaches, combined with other commonly used systems such as chemical or electro methods, may lead to future hybrid systems.

Dan Luo and W. Mark Saltzman

Literature Cited

1. Luo D, Saltzman WM. Nat Biotechnol 2000; 18(1):33-37.
2. Huang L, Hung MC, Wagner E. *Non-Viral Vectors for Gene Therapy*, 1st edn., Academic Press, New York, 1999.

Chapter 1

Supramolecular Assemblies of DNA Delivery Systems

Bruno Pitard

Abstract

Gene delivery into cultured cells or *in vivo* is a promising approach to the treatment of diseases. Several gene delivery systems have been developed to promote gene expression either *in vitro* or *in vivo*. Concerns about viral-induced immune responses, the risk associated with replication-competent viruses, and production issues have stimulated efforts toward the development of alternative gene delivery systems such as cationic lipids and polymers. These positively charged molecules interact through electrostatic forces with DNA. This results in the formation of highly organized supramolecular structures where DNA molecules are condensed and protected against DNAses degradation. Association of DNA with cationic lipids under a micellar or liposomal form leads to lamellar organization with DNA molecules sandwiched between lipid bilayers. Although the lamellar phase is the common described structure, as evidenced by small-angle X-ray scattering and electron microscopy, monovalent cationic lipid combined with a hexagonal forming lipid resulted with DNA in an inverted hexagonal structure. Despite a lot of effort, the mechanism of gene transfer with cationic carrier is still ill-defined. Therefore, correlations need to be established between physicochemical properties of synthetic DNA delivery systems and *in vitro* and *in vivo* transfection efficiency.

Introduction

DNA delivery systems are used in the field of gene therapy to introduce foreign DNA encoding therapeutic protein sequences into cells. Ideally, these DNA delivery systems should

1. be safe, nontoxic, nonimmunogenic, and well tolerated
2. contain DNA in small particles; below 500 nm in diameter in order to be compatible with cell penetration through endocytosis
3. provide a stealthy behavior in order to improve in vivo bioavailability; i.e. prevent the removal of particles from the blood by the monophagocytic system

---

Abbreviations: cmc, critical micellization concentration; DODAB, dioctadecyldiammonium bromide; DOTAP, dioleoyl trimethylammonium propane; BGTC, bisguanidinium tren cholesterol; EDOPC, 1,2-dioleoyl-sn3-ethylphosphocholine; DOPE, dioleoylphosphatidylethanolamine; DOPC, dioleoylphosphatidylcholine.

4. target particular cells
5. penetrate the target cell
6. escape from the endosomes
7. protect DNA molecules from extracellular and intracellular DNAse degradation
8. target the nucleus for the transgene expression

The future of gene therapy will depend on the ability to engineer all the properties listed above into DNA delivery systems. For this reason both viral and nonviral delivery systems have been investigated towards the development of effective gene delivery systems. Nonviral vectors result essentially from the association of DNA with cationic molecules such as lipids and polymers. Cationic lipid/DNA and cationic polymer/DNA complexes are called lipoplexes and polyplexes, respectively. Despite the extensive use and diverse applications of lipoplexes or polyplexes, it must be stressed that their surpramolecular characteristics and their mechanisms of transfection are in many ways surprisingly ill-defined. This is to some extent understandable, as most of the previous efforts were focused on the synthesis of new cationic lipids or polymers and not on the understanding of physicochemical parameters involved in the formation and transfection of lipoplexes. In this paper we present the physicochemical characteristics of nonviral vectors for two length scales: macroscopic and microscopic. We also present the relationships between their physicochemical properties and the transfection efficiency.

Lipoplexes on a Macroscopic Length Scale

Lipoplexes are formed by mixing plasmid DNA with cationic lipids under a liposomal or micellar form. Mixing polyanionic DNA with the polycationic surface of the liposomes or micelles leads to the formation of lipoplexes through electrostatic interaction between the two oppositely charged molecules. The lipoplexes are usually described by the theoretical charge ratio (expressed as moles of positive charge per mole of negative charge). Cationic lipids (Fig. 1.1) are composed of a hydrophobic moiety (cholesterol, acyl chains, etc.) linked to a cationic headgroup including quaternary ammonium such as dioctadecyldiammonium bromide (DODAB), or dioleoyl trimethylammonium propane (DOTAP),[1] or a polycation such as a lipopolyamine (RPR120535)[2] or bisguanidinium tren cholesterol (BGTC).[3] During the formation of lipoplexes, DNA is complexed and the initial lipid organization is rearranged to form supramolecular assemblies. Two essential

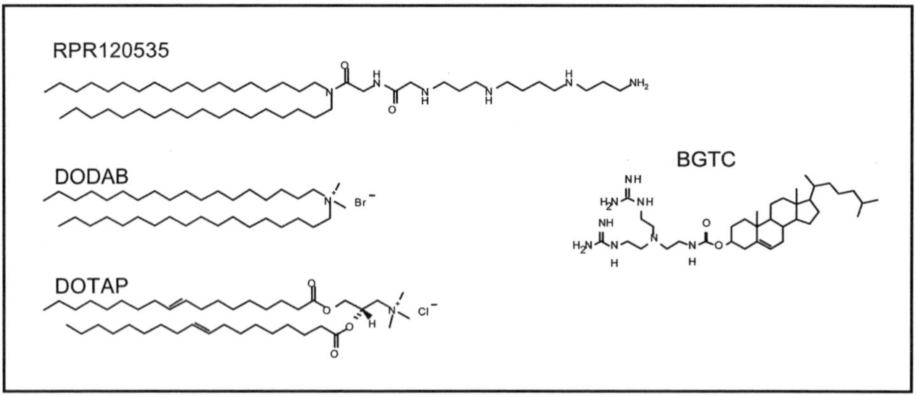

Fig. 1.1. Structure of the cationic lipids lipopolyamine RPR120535, dioctadecyldiammonium bromide (DODAB), dioleoyltrimethylammonium propane (DOTAP), and bisguanidinium tren cholesterol (BGTC).

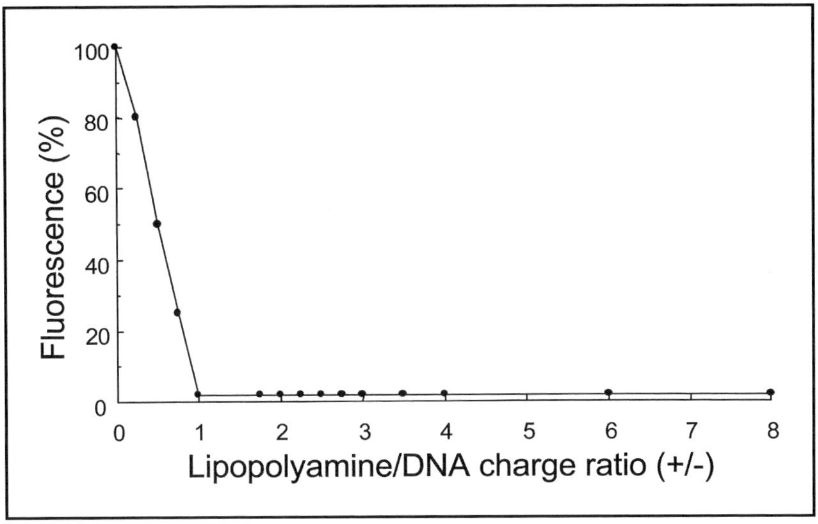

Fig. 1.2. DNA entrapment as a function of the cationic lipid/DNA ratio. To assess DNA entrapment, ethidium bromide was added to lipoplex solutions 1 h after complexation at room temperature, and fluorescence was monitored. For lipopolyamine RPR120535/DNA ratios from 0 to 1, fluorescence intensity sharply decreased from 100% to a value close to 0 and stayed at this minimal level for charge ratios above 1.

physicochemical properties for in vitro or in vivo applications can be monitored at the macroscopic length scale, i.e. DNA complexation and the size of the lipoplexes.

DNA Complexation

Fluorescence experiments are usually performed to evaluate DNA condensation in the formed aggregates.[4–6] Thus, lipoplexes are exposed to ethidium bromide molecules, which, upon intercalation between DNA base pairs, act as a fluorescence probe. Figure 1.2 shows a typical experiment, with the lipopolyamine RPR120535, where fluorescence intensity is monitored as a function of the theoretical charge ratio, which is expressed as moles of positive charge per mole of negative charge. For charge ratios ranging from 0 to 1, fluorescence intensity sharply decreases, from 100% to a value close to zero and stays at this level for lipoplexes with a charge ratio above 1. This indicates that plasmid DNA was, in part, accessible to the probe for charge ratios ranging from 0 to 1, whereas for higher charge ratios, DNA molecules were condensed by the cationic lipid. This observation is common to several cationic lipids.

Gel electrophoresis on a 0.8% agarose gel stained with ethidium bromide is also used to follow DNA condensation by cationic molecules.[5,6] Figure 1.3 shows that the amount of uncomplexed or free DNA decreases as the charge ratio (+/−) is raised. A charge ratio in excess of 1 resulted in aggregates that did not migrate out of the well under these conditions. This observation confirms the above spectrofluorescence experiments performed in solution.

Colloidal Stability and the Size of Lipoplexes

Colloidal stability and the size of lipoplexes are key parameters for the in vivo application of nonviral gene delivery systems. To assess the colloidal stability of nonviral vectors,

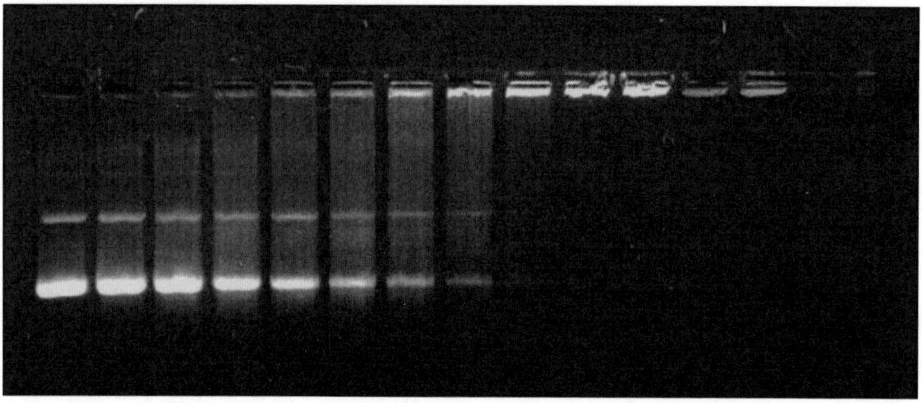

Fig. 1.3. Agarose gel electrophoresis of lipopolyamine RPR120535/DNA lipoplexes formed at different charge ratios. Wells contain lipid and DNA at the following charge ratio (left to right): 0, 0.1, 0.2, 0.3, 0.4, 0.5, 0.6, 0.7, 0.9, 1, 2, 4, and 8. Agarose gel electrophoresis was used to assess the relative amounts of DNA that were free or incorporated into lipoplex as a function of the lipid/DNA ratio. DNA complexed with lipid did not migrate out of the well under these conditions because of the size of lipoplexes. The amount of free DNA decreased as the charge ratio was raised.

dynamic light scattering was performed after 1 h of complexation at room temperature. This time period was used to take into account the time necessary for the delivery of the lipoplexes into a biological host. Figure 1.4 illustrates the evolution of the mean diameter as a function of the charge ratio. Colloidal stability of lipoplexes could be defined by three main zones (A, B, and C). Zone A was comprised of colloidally stable complexes with a negative charge ratio and uncondensed DNA surrounding the lipoplexes. Zone B complexes were essentially neutrally charged with condensed DNA contained within colloidally unstable aggregates. Finally, zone C lipoplexes were composed of colloidally stable particles with condensed DNA and positive charge ratios, thus strongly repelling each other. Modulation of the lipoplex formulations had a profound effect on the zone of colloidal stability obtained for a given charge ratio. For example, increasing of the NaCl concentration, the cationic lipid acyl chain length, the pH, and the DNA concentration extended zone B (aggregated lipoplexes), while decreasing zone C (colloidally stable lipoplexes). Consequently, for the same charge ratio of two different formulations, the resultant lipoplexes could be of two completely different colloidal stabilities, revealing that the charge ratio alone is not the final determinant of the colloidal state of lipoplexes.[7]

One of the advantages of nonviral delivery systems over viral vectors is the possibility to transfect cells with large DNA molecules. Investigations on the effect of plasmid DNA size on colloidal stability of lipoplexes showed that plasmid size from 900 to 52,500 bp does not affect either the morphology or the structure of cationic lipid/DNA lipoplexes.[8] Therefore, the number of plasmid molecules per lipoplexes was inversely proportional to the DNA length, i.e. the largest number of plasmid DNA molecules per lipoplex is obtained with the smallest plasmid.

Recently, the laboratory of J.P. Behr has proposed a strategy that enables them to entrap only one DNA molecule per lipoplex.[9] This strategy is based on the use of oligomerizable cationic detergent with one acyl chain. DNA condensation occurs with monomer of cationic detergent (below the critical micellization concentration (cmc)), and then the organization of the lipoplexes is frozen by linking cationic detergent molecules

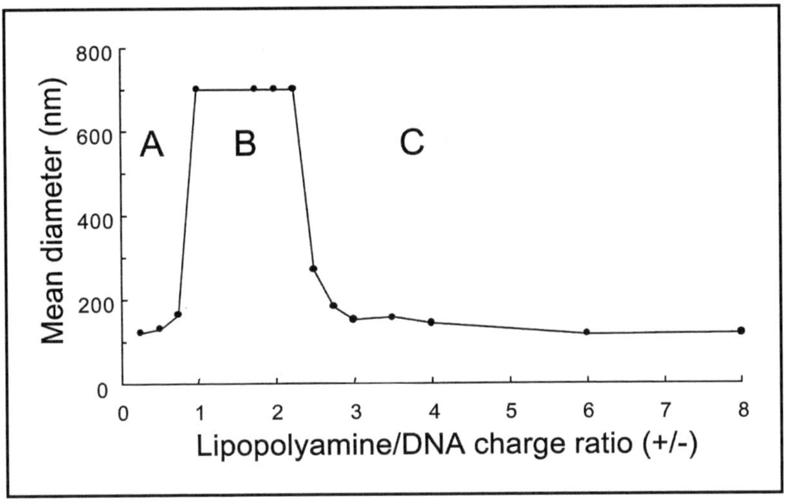

Fig. 1.4. Colloidal stability of lipopolyamine RPR120535/DNA lipoplexes as a function of the charge ratio. To assess the colloidal stability of lipoplexes, dynamic light scattering analysis was performed after 1 h of complexation at room temperature. Lipoplexes from zones A and C were stable, whereas those from zone B exhibited a visible precipitate. Thus, an arbitrary value of 700 nm was attributed to the zone B lipoplex particle size because the diameter of the precipitated lipoplexes cannot be measured by dynamic light scattering.

via disulfide bonds. The resulting lipoplexes have a mean diameter of 23 nm, a negative zeta potential, and are stable in physiological conditions.

Lipoplexes on a Microscopic Length Scale

Structure of Lipoplexes

Small-angle X-ray scattering(SAXS) has been used to probe the structure of cationic lipid/DNA lipoplexes at the microscopic length scale. Table 1.1 summarizes lipoplexes structure revealed by this technique. It was shown by several research groups that highly ordered phases are formed when DNA molecules are mixed with cationic liposomes or micelles. Cationic liposomes composed of different cationic lipids such as DOTAP,[10] DODAB,[11,13] EDOPC (1,2-dioleoyl-sn3-ethylphosphocholine),[14] BGTC,[15] and DOGS (dioactadecylamidoglycylspermine)[16] exhibited multilamellar phases with DNA molecules sandwiched between cationic lipid bilayers. The periodicity of the lamellar organization depends on the cationic liposomes used (Table 1.1). In the case of DOTAP/DOPC,[10] DODAB/DOPE,[13] and DODAB/cholesterol[11] (DOPC, dioleoylphosphatidylcholine; DOPE, dioleoylphosphatidylethanolamine) a second repeat distance was observed and was attributed to the organization of the DNA molecules in a 2D plane between lipid bilayers. This periodicity of DNA packing within the lipid bilayers ranged from 2.45 to 5.71 nm, as the neutral lipid DOPC increased for a constant DOTAP/DNA charge ratio.[10] Recently, inverted hexagonal phases (Fig. 1.5) were observed with lipoplexes resulting from the association of DNA with cationic liposomes containing a high content of DOPE, a hexagonal forming lipid.[12,13] In this case, DNA molecules are

Table 1.1. Analysis by SAXS of the structure of lipoplexes resulting from the association of DNA with various cationic lipids under different morphologies (i.e. liposomes, spherical, or thread-like micelles)

Cationic lipid morphology[a]	Structure	Periodicity[b]	Reference
Cationic liposomes			
DOTAP/DOPC (1/1)	Multilamellar structure	$d_{\text{lipid}} = 65.1$ Å $d_{\text{DNA}} = 24.5$–57.1 Å	10
DODAB/Chol (1/1)	Multilamellar structure	$d_{\text{lipid}} = 64.4$ Å $d_{\text{DNA}} = 36$ Å	11
DOTAP/DOPE	Multilamellar and inverted hexagonal structure depending on the DOTAP/DOPE molar ratio	*TAP/PE < 1/0.65: L_α^c ($d_{\text{lipid}} L_\alpha^c = 63.47$ Å; $d_{\text{DNA}} = 36.5$ Å) *1/0.65 < TAP/PE < 1/2.19: L_α^c & H_{II}^C (d_{lipid} $L_\alpha^c = 63.47$ Å; $d$lipid $H_{II}^C = 67.4$ Å) *1/2.19 < TAP/PE < 1/5.32: H_{II}^C (d_{lipid} $H_{II}^C = 67.4$ Å) *TAP/PE > 1/5.32: H_{II}^C & H_{II} (d_{lipid} $H_{II}^C = 65.1$ Å; d_{lipid} $H_{II} = 74.41$ Å)	12
DODAB/DOPE	Multilamellar and inverted hexagonal structure depending on the DODAB/DOPE molar ratio	*DAB/PE = 3/1: L_α^c (d_{lipid} $L_\alpha^c = 64.8$ Å; $d_{\text{DNA}} = 36.1$ Å) *DAB/PE = 2/3: L_α^c & H_{II}^C (d_{lipid} $L_\alpha^c = 64.8$ Å; d_{lipid} $H_{II}^C = 66.7$ Å) *DAB/PE = 1/3: H_{II}^C (d_{lipid} $H_{II}^C = 68.2$ Å)	13
EDOPC	Multilamellar structure	$d_{\text{lipid}} = 62.5$ Å	14
BGTC/DOPE (3/2)	Multilamellar structure	$d_{\text{lipid}} = 70$ Å	15
DOGS	Multilamellar structure	$d_{\text{lipid}} = 65$ Å	16
DOGS/DOPE/cardiolipin (0.35/0.52/0.13)	Multilamellar structure	$d_{\text{lipid}} = 69$ Å	17
Cationic micelles			
Spherical micelles RPR120535	Multilamellar structure	$d_{\text{lipid}} = 80$ Å	18
Thread-like micelles			
BGTC	Multilamellar structure	$d_{\text{lipid}} = 70$ Å	15

[a] d_{lipid} and d_{DNA} represent respectively the periodicity of the lamellar organisation of lipids and the 2D arrangement of DNA molecules in the compaction plane. L_α^c and H_{II}^C represent respectively the lamellar and the inverted hexagonal phase of DOTAP/DOPE-DNA or DODAB/DOPE-DNA lipoplexes. H_{II} is the hexagonal phase of pure DOPE.

[b] DOTAP, dioleoyl trimethylammonium propane; DODAB, dioctadecyldiammonium bromide; EDOPC, 1,2-dioleoyl-sn3-ethylphosphocholine; DOPE, dioleoylphosphatidylethanolamine; BGTC, bisguanidinium tren cholesterol; DOGS, dioactadecylamidoglycylspermine. DOTAP/DOPE (TAP/PE) and DODAB/DOPE (DAB/PE) are both expressed as a molar ratio.

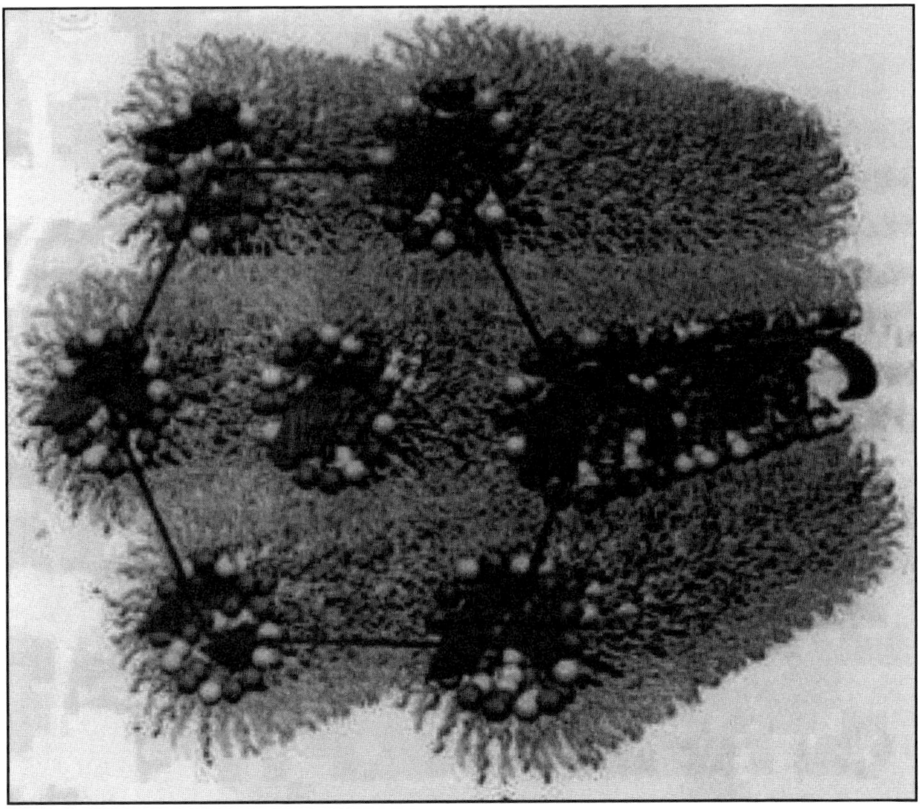

Fig. 1.5. Schematic representation of the inverted hexagonal phase of DOTAP–DOPE/DNA lipoplexes. DNA molecules are surrounded by lipid monolayers of DOTAP/DOPC and arranged on a 2D hexagonal lattice. Reprinted (abstract/excerpted) with permission from Ref. 16, p. 78. Copyright 1998 American Association for the Advancement of Science.

coated with cationic lipid monolayers and arranged on a 2D hexagonal lattice. However, other formulations involving DOPE as a colipid such as BGTC/DOPE–DNA did not exhibit a hexagonal structure even if the amount of DOPE was increased.[15] This discrepancy may be explained by the significant differences between DOTAP or DODAB and BGTC. Indeed, DOTAP and DODAB are monovalent cationic lipids with two acyl chains whereas BGTC bears two positive charges linked to an hydrophobic moiety consisting of cholesterol. This highlights the absence of general rules regarding the effect of DOPE on the formation of the hexagonal structure.

Complexation of plasmid DNA with spherical micelles of about 5 nm in diameter, instead of cationic liposomes, leads to the formation of smaller lipoplexes consisting of ordered lamellar domains characterized by an even spacing of 8 nm irrespective of the cationic lipid/DNA ratio.[18] Indeed, X-ray scattering scans and micrographs of negatively charged (zone A, Fig. 1.6), neutral (zone B, Fig. 1.6), and positively charged lipoplexes (zone C, Fig. 1.6) exhibited lamellar organization. Therefore, plasmid DNA is able to transform multivalent cationic lipid micelles into a supramolecular organization characterized by local lamellar microdomains (Fig. 1.6).

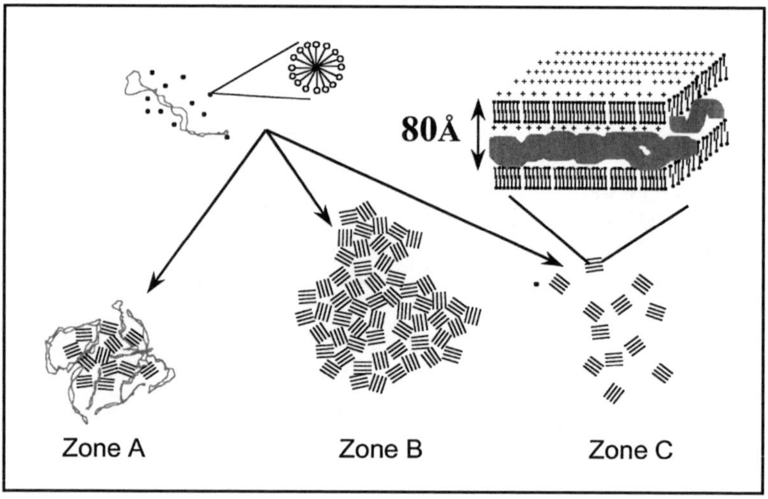

Fig. 1.6. Schematic representation of lipoplexes resulting from the association of plasmid DNA with spherical cationic micelles of 5 nm in diameter. Structural analysis, by electron microscopy and small-angle X-ray scattering, showed that lipoplexes morphology depends on the charge ratio, but the presence of lamellar microdomains is observed irrespective of the charge ratio. A, B, and C represent negatively, neutrally, and positively charged lipoplexes, respectively. From Ref. 17. Copyright (1997) National Academy of Sciences, U.S.A.

Morphology of Lipoplexes

Data obtained from X-ray scattering experiments are extremely useful in understanding the inner structure of cationic lipid/DNA lipoplexes at the molecular level. Remarkably, the results have shown that cationic lipid/DNA lipoplexes resulting from the association of either micelles or liposomes composed of mono-or polyvalent cationic lipid usually give rise to multilamellar complexes of various even spacing depending on the cationic lipid molecule used (Table 1.1). DNA molecules in this multilamellar structure are sandwiched between lipid bilayers. Another structure of lipoplexes was also observed: an inverted hexagonal phase. This structure was generated with the cationic liposomes of DOTAP/DOPE with a molar ratio ranging from 1/2.19 to 1/5.32 and with DODAB/DOPE (1/3 M/M) liposomes.

However, the structural organization of lipoplexes provided by SAXS is the mean structure present in the sample. At the same time minor structures, if existing, could not be detected. For example, it was recently suggested that both multilamellar and hexagonal phases could coexist in the same particle, even though an X-ray scattering scan indicated a multilamellar structure.[17] Moreover, no information on the morphology of lipoplexes is provided by SAXS. Therefore, research groups used electron microscopy techniques to visualize the morphology of lipoplexes, to understand the mechanism of the formation of lipoplexes and to attribute the peaks observed by X-ray scattering. Various forms of electron microscopy have been used to study cationic lipid/DNA lipoplexes of different lipid compositions, and different size and topological features of the DNA molecules, e.g. supercoiled plasmid and linear DNA. Freeze-fracture electron microscopy has shown uncondensed DNA fibers coated with lipid connected to the surface of aggregated cationic liposomes of 3-β-(*N*-(*N'*,*N'*-dimethylethane)carbamoyl) cholesterol/DOPE (DC-Chol/DOPE).[19] Electron microscopy performed with negative staining of DNA complexed with the cationic liposome of *N*-(1-(2,3-bis(oleoyloxy))propyl)-*N*, *N*, *N*-trimethylammonium chloride/DOPE (DOTMA/DOPE) suggested that cationic liposomes initially form clusters along the DNA molecule. Above a critical liposome density, these clusters coalesce by DNA-induced membrane fusion, leading to encapsulation of DNA molecules by the lipids.[20] Transmission electron microscopy (TEM) with uranyl acetate staining further indicated that DNA molecules might be condensed and intertwined in the tubules of the lipopolyamine DOGS network, leading to hexagonal packing of DNA.[21] Cryo-transmission electron microscopy studies of lipoplexes formed from various cationic lipids were performed by several research groups.[11,13,15,18,22–25] Cryo-electron micrographs of lipoplexes showed ordered multilamellar particles characterized by an even spacing. This spacing was measured routinely by optical diffraction, which supported the interlamellar spacing found by SAXS. The interlamellar spacing found in these studies ranged from 6.5 to 8.0 nm depending on the cationic lipid molecule used. The number of lamellae contained per multilamellar structure observed by cryo-electron microscopy was in good agreement with those calculated from the half-width and the position of the main X-ray scattering peak of the lamellar symmetry. However, in these reported studies, DNA molecules were either invisible or barely visible within the lipoplexes structures. Thus, the short periodicity, interpreted to be the DNA interaxial spacing of the 1D array of aligned DNA sandwiched between lipid bilayers, could not be definitely detected. In some studies, however, DNA molecules could be visualized by cryo-TEM wrapping around cationic liposomes with a periodicity compatible with those observed by SAXS.[25,26]

Physicochemical–Transfection Efficiency Relationships

Ever since the synthesis of the first cationic lipid by Philip Felgner and his collaborators,[27] a variety of cationic lipids for gene transfer have been synthesized.[28,29] They can be classified into two different groups: (i) monovalent cationic lipids containing only one positive charge, and (ii) polyvalent cationic lipids in which the lipid moiety is usually linked through a spacer to a polyamine. The main efforts have been dedicated to generating modifications of the lipid chain and of the cationic polar headgroup such as geometrically differing polyamines.[2] These cationic lipids are then assessed for their in vitro transfection efficiency. Therefore, there is still a limited understanding of the factors and processes leading to transfection efficiency because cationic lipids are designed on the basis of trial and error, using expression of the transgene as the end point. However, a better understanding of the lipoplex structures involved in their biological activity, cellular entry, and nuclear translocation of plasmid DNA might improve their efficiency for transfection. In particular, correlations need to be established between the supramolecular assemblies formed by the lipoplexes and their subsequent transfection efficiency either in vitro or in vivo.

Hexagonal Packing of DNA

In a recent investigation,[30] the luciferase activity measured in transfected mouse fibroblast L-cells showed that the inverted hexagonal structure (Fig. 1.6) resulting from the association of DNA with cationic liposomes of DOTAP/DOPE (1/2.41 M/M), was by far the most active in promoting transfection. The transfection activity of these lipoplexes was 3 log greater than the activity of the DNA complexes formed with cationic liposomes of DOTAP/DOPC (1/2.28 M/M) consisting of a lamellar structure. Experiments were conducted to study the intracellular trafficking of these two types of supramolecular organizations. The results suggested that the difference in transfection efficiency is a result of differences in intracellular behavior, because self-assembled lamellar lipoplexes that were internalized into cells were stable, whereas hexagonal complexes led to aggregated clusters that fused with the cell membrane, thus facilitating DNA escape.

Small DNA Molecules

An important stage in gene therapy in the future will be the transfection of large DNA fragments containing large transgene with regulatory systems for gene expression induced by chemical agents. Thus it will be of crucial importance to develop nonviral delivery systems capable of transducing large DNA molecules into cells. Structural analysis of the complexes resulting from the association of plasmid DNA of various sizes, ranging from 900 to 52,500 bp with cationic lipids, showed that lipoplex morphology and structure do not depend on DNA length. The periodicity of the lamellar structure is determined by the cationic lipid, and the spherical morphology by the charge density of the DNA molecule.[8] On the other hand, transfection efficiency was affected by the size of the plasmid DNA molecule, i.e., small DNA molecules had higher transfection efficiencies than larger ones. This may be due to the limited diffusion of large DNA molecules from the endosomal compartment to the cytosol and to the nuclear pore complex into the nucleus.[31]

Large-Sized Lipoplexes

Lipoplex size and stability were also shown to be key parameters affecting transfection efficiency. In particular, large-sized, colloidally unstable lipoplexes with a mean diameter over 700 nm induced efficient transfection in the presence or in the absence of serum.[7,32] On the other hand, colloidally stable complexes of less than 200 nm in diameter led to efficient transfection only in the absence of serum. These colloidally stable or unstable lipoplexes could be obtained by varying the reaction mixture conditions such as NaCl concentration, pH, or DNA concentration. The clear correlation between large-sized lipoplexes and high levels of in vitro transgene expression was shown to be independent of both the percentage of DNA internalized into cells and the lamellar organization of the cationic lipid/DNA lipoplexes.

Mechanism of DNA Release inside the Cells

It has been proposed that DNA dissociation from cationic lipid and its subsequent release into the cytoplasm is triggered by the electrostatic interaction between cationic lipids from the lipoplexes and anionic lipids from the endosomal membrane.[33] This results in a DNA displacement from the cationic lipids and led to the release of DNA into the cytoplasm. DNA dissociation from the cationic lipids is of crucial importance because it has been shown that DNA molecules have to be free of any cationic vectors in order to be expressed.[34,35]

Conclusion

The understanding of the physicochemical characteristics of lipoplexes involved in the in vitro transfection efficiency is increasing. However, there is still a limited understanding of the behavior of lipoplexes in vivo and inside cells. In particular, correlations need to be established between the supramolecular assemblies formed by lipoplexes and all the steps that lead to transfection.

Acknowledgments

I am grateful to my colleagues in Aventis Pharma and Pasteur, and Rhodia for fruitful discussions and collaborations.

Literature Cited

1. Stamatatos L, Leventis R, Zuckerman MJ, Silvius JR. Biochemistry 1988; 27:3917-3925.
2. Byk G, Dubertret C, Escriou V, Frederic M, Jaslin G, Rangara R, Pitard B, Crouzet P, Wils P, Schwartz B, Scherman D. J Med Chem 1998; 41:224-235.
3. Vigneron JP, Oudrhiri N, Fauquet M, Vergely L, Bradley JC, Basseville M, Lehn P, Lehn JM. Proc Natl Acad Sci USA 1996; 93:9682-9686.
4. Ferrari M, Nguyen CM, Zelphati O, Tsai Y, Felgner P. Hum Gene Ther 1998; 9:341-351.
5. Eastman SJ, Siegel C, Tousignant J, Smith AE, Cheng SH, Scheule RK. Biochim Biophys Acta 1997; 1325:41-62.
6. Bhattacharya S, Mandal S. Biochim Biophys Acta 1997; 1323:29-44.
7. Turek J, Dubertret C, Jaslin G, Antonakis K, Scherman D, Pitard B. J Gene Med 2000; 2:32-40.
8. Kreiss P, Cameron B, Rangara R, Mailhe P, Aguerre O, Airiau M, Scherman D, Crouzet J, Pitard B. Nucleic Acids Res 1999; 27:3792-3798.
9. Blessing T, Remy JS, Behr JP. Proc Natl Acad Sci USA 1998; 95:1427-1431.
10. Raedler JO, Koltover I, Salditt T, Safinya CR. Science 1997; 275:810-814.
11. Lasic D, Strey H, Stuart M , Podgornik R, Frederik PM. J Am Chem Soc 1997; 119:832-833.
12. Koltover I, Salditt T, Rädler JO, Safinya CR. Science 1998; 281:78-81.
13. Mel'nikova Y, Mel'nikov SM, Löfroth JE. Biophys Chem 1999; 81:125-141.
14. MacDonald RC, Ashley GW, Shida MM, Rakhmanova VA, Tarahovsky YS, Pantazatos DP, Kennedy MT, Pozharski EV, Baker KA, Jones RD, Rosenzweig HS, Choi KL, Qiu R, Mcintosh TJ. Biophys J 1999; 77:2612-2629.
15. Pitard B, Oudrhiri N, Vigneron JP, Hauchecorne M, Aguerre O, Toury R, Airiau M, Ramasawmy R, Scherman D, Crouzet J, Lehn JM, Lehn P. Proc Natl Acad Sci USA 1999; 96:2621-2626.
16. Boukhnikachvili T, Aguerre-Chariol O, Airiau M, Lesieur S, Ollivon M, Vacus J. FEBS Lett 1997; 409:188-194.
17. Schmutz M, Durand D, Debin A, Palvadeau, Y, Etienne A, Thierry AR. Proc Natl Acad Sci USA 1999; 96:12293-12298.
18. Pitard B, Aguerre O, Airiau M, Lachagés AM, Boukhnikachvili T, Byk G, Dubertret C, Herviou C, Scherman D, Mayaux JM, Crouzet J. Proc Natl Acad Sci USA 1997; 94:14412-14417.
19. Sternberg B, Sorgi F, Huang L. FEBS Lett 1994; 3356:361-366.
20. Gershon H, Ghirlando R, Guttman S, Minsky A. Biochemistry 1993; 32:7143-7151.
21. Labat-Moleur F, Steffan AM, Brisson C, Perron H, Feugas O, Furstenberger P, Oberling F, Brambilla E, Behr J-P. Gene Ther 1996; 3:1010-1017.
22. Gustafsson J, Arvidson G, Karlsson G, Almgren M. Biochim Biophys Acta 1995; 1235:305-312.
23. Templeton N, Lasic D, Frederik P, Strey H, Roberts D, Pavlakis G. Nat Biotechnol 1997; 15:647-652.
24. Battersby BJ, Grimm R, Huebner S, Cevc G. Biochim Biophys Acta 1998; 1372:379-383.
25. Hubner S, Battersby BJ, Grimm R, Cevc G. Biophys J 1999; 76:3158-3166.
26. Pitard B, Oudhiri N, Vigneron JP, Lambert O, Vivien E, Hauchecorne M, Scherman D, Rigaud JL, Lehn JM, Lehn P. J Gene Med 2001; 3:478-487.
27. Felgner PL, Gadek TR, Holm M, Roman R, Chan HW, Wenz M, Northrop JP, Ringold GM, Danielsen M. Proc Natl Acad Sci USA 1987; 84:7413-7417.
28. Behr JP, Demeneix B, Loeffler JP, Perez-Mutul J. Proc Natl Acad Sci USA 1989; 86:6982-6986.
29. Gao XA, Huang L. Biochem Biophys Res Commun 1991; 179:280-285.
30. Lin AJ, Slack NL, Ahmad A, Koltover I, George CX, Samuel CE, Safinya CR. J Drug Target 2000; 8:13-27.
31. Lukacs GL, Haggie P, Seksek O, Lechardeur D, Freedman N, Verkman AS. J Biol Chem 2000; 275:1625-1629.
32. Ross PC, Hui SW. Gene Ther 1999; 6:651-659.
33. Xu Y, Szoka FC. Biochemistry 1996; 35:5616-5623.
34. Zabner J, Fasbender AA, Moninger T, Poellinger KA, Welsh MJ. J Biol Chem 1995; 270:18997-19007.
35. Pollard H, Remy JS, Loussouarn G, Demolombe S, Behr JP, Escande D. J Biol Chem 1998; 273:7507-11.

CHAPTER 2

Engineering the Surface Properties of Synthetic Gene Delivery Vectors

Gary Lee and David Schaffer

Abstract

Synthetic gene delivery vehicles are a highly promising approach to gene delivery; however, several problems must still be overcome before they can begin to enjoy clinical success. A number of these problems can be addressed by engineering and optimizing the properties of the vector surface, the component of the particle that interacts and "communicates" with tissues and cells during the delivery process. Surfaces must be engineered to satisfy two ostensibly conflicting constraints: the ability to interact specifically with a target cell while avoiding nonspecific protein interactions, particularly with components of the immune system. We summarize progress that has been made in both these areas and discuss several approaches where the intersection of biological and chemical solutions promises to significantly advance the engineering of synthetic vehicles.

Introduction

Synthetic vectors enjoy several advantages over viral vehicles, among them a large degree of flexibility in the design of their properties. A wide variety of materials have been employed for nucleic acid condensation, controlled release, surface engineering and cell targeting, and enhancement of delivery efficiency. In addition, alternative methods for assembling the same materials into a vector can yield structures with widely different properties. Faced with such an enormous number of choices in constructing synthetic vehicles, however, this design flexibility rapidly turns from an advantage into a daunting problem: how does one identify regions of this vast parameter space that optimize vector function?

One approach that has guided the development of synthetic vectors since their inception is the imitation of viruses. After tens of millions of years of trial and error, viruses have evolved numerous strategies to deliver their genetic payload to target cells efficiently. By chemically synthesizing materials that mimic these activities and properties of viruses, or even by directly borrowing materials from viruses, synthetic vector performance has been significantly improved, and the region of their parameter space that must be searched before success is achieved has been narrowed. The continuing synergy between newly developing chemical and biological approaches to this problem may advance the field of synthetic vectors until their capabilities become comparable to those of viruses.

Synthetic DNA Delivery Systems, edited by Dan Luo and W. Mark Saltzman. © 2003 Eurekah.com. and Kluwer Academic/Plenum Publishers.

One particular area of work where chemistry and biology approaches have collaborated is in the design of the surface properties of synthetic vectors. It is the vector surface that first interacts and "communicates" with cells and tissues in vivo, and it thereby determines the success of the vehicle from the time of injection in vivo to its entry into a target cell. Surfaces must simultaneously be biologically inert to evade neutralization by the immune system, and biologically active to dock with cells targeted for gene delivery. We will discuss continuing approaches towards the design of synthetic vectors that satisfy these two, often conflicting demands.

Targeted Synthetic Vectors

Background and Recent Advances

Synthetic vector particles are constructed by complexing negatively charged nucleic acids with synthetic materials, most often positively charged, such as lipids to generate liposomes (lipoplexes) or polymers to generate molecular conjugates (polyplexes). The resulting particles dock with the cell surface by two mechanisms. First, if the net vector charge is positive, they nonspecifically bind to the negatively charged cell membrane, particularly to proteoglycans.[1] Since proteoglycans are ubiquitous, the resulting delivery is very nonspecific. However, delivering genes to specific cell populations in vivo is often desired to reduce toxicity, other side effects, or the required vector dosage. Therefore, significant effort has been invested to develop vectors capable of targeting gene delivery to specific cells.[2]

In general, a targeted vector is created by examining the known repertoire of cell surface proteins expressed by the target cell, identifying one receptor that is selectively expressed by that specific cell type, choosing a ligand that binds the receptor, and linking the ligand to the vector surface. This ligand then emulates viral attachment proteins in mediating virion binding and cellular entry. For example, the first molecular conjugates developed by Wu and colleagues employed asialoglycoproteins chemically crosslinked to the polycation in order to mediate specific delivery in vitro and in vivo, albeit at a relatively low efficiency, to hepatocytes via the asialoglycoprotein receptor.[3,4] In parallel, Huang et al. used an antibody to target pH-sensitive liposomes, or immunoliposomes, for gene delivery to a tumor model in vivo.[5]

Work in the years since these pioneering studies has repeatedly demonstrated that a large variety of ligands, as well as several of methods of attaching them to the vehicle, can be utilized successfully. The types of targeting molecules employed fall into several categories. First, one very promising group of cell surface receptor targets are the integrin cell adhesion receptors. These heterodimeric proteins, composed of an α and a β integrin subunit, bind ligands such as collagen or laminin that contain an arginine-glycine-aspartic acid (RGD)* motif. However, the expression of particular $\alpha\beta$ integrin combinations is often localized to specific cell or tissue types, and each heterodimer has the potential to bind to an RGD sequence presented in a specific context or conformation. Therefore, the use of RGD-containing peptides to target integrin receptors is an active and promising area of research. For example, RGD peptides have been employed for targeted gene delivery with polylysine or polyethylenimine molecular conjugates[6,7] as well as with liposomes.[8]

* Abbreviations used in this chapter are as follows: RGD, arginine-glycine-aspartic acid motif; PEG, polyethylene glycol; HIV, human immunodeficiency virus.

These vehicles conceptually mimic viruses such as adenovirus that utilize integrin as a receptor.[9]

Carbohydrates are another class of ligands that offer the potential for targeted delivery, particularly to hepatocytes and cells of the immune system. As discussed above, desialyated glycoproteins can mediate delivery to hepatocytes.[3,4] In addition, crosslinking small monosaccharides or oligosaccharides to gene delivery vehicles offers the potential for specific delivery to cells expressing other lectins, receptors that bind sugars.[10] For example, Hanson, Ferkol, and colleagues have repeatedly demonstrated the potential of monosaccharides in targeted gene delivery. They have used mannose chemically crosslinked to polylysine or polyethylenimine molecular conjugates to target delivery to macrophages[11,12] and dendritic cells[13,14] in vitro or in vivo.

Several gene delivery strategies exploit receptors that function to carry nutritional compounds into cells. For example, the transferrin receptor, which serves to import iron into a cell, was one of the first targets for molecular conjugate gene delivery.[15] In addition, the folate receptor, believed to function in the cellular uptake of the metabolite folate, has been targeted using both its natural folate ligand as well as antibodies against the receptor.[16] In addition, since the receptors for folate and transferrin are overexpressed in a variety of tumors, targeted delivery with these ligands may prove effective for cancer gene therapy.

In addition to nutritional receptors, delivery can be targeted to signaling receptors, whose potentially more restricted expression by specific cell types offers promise for targeted delivery. For example, the fibroblast growth factor receptor has been targeted for gene delivery in culture.[17,18] In addition, vectors displaying the ligand EGF have been used to target the epidermal growth factor receptor in vitro.[19,20] Although a large number of cell types express this receptor in vivo, its overexpression by some tumors could be exploited for cancer gene therapy. As a final example, the neuropeptide neurotensin was recently crosslinked to polylysine for delivery to neural cell lines.[21]

The studies above exploit natural ligands to target delivery to receptors of interest. However, antibodies provide a general alternative for targeted delivery via surface antigens, even if no natural ligand is known. In fact, antibodies have been the most commonly used ligand for targeted liposome delivery[5] and are particularly promising for targeted delivery via cell surface, tumor-specific antigens in cancer gene therapy. For example, Mohr et al. achieve targeted gene delivery by linking lipoplexes to an antibody against an uncharacterized surface glycoprotein overexpressed by hepatocellular carcinoma and other cells.[22] An antibody was also used for targeted delivery to lung endothelial cells in vivo via the platelet endothelial cell adhesion molecule.[23] In a final example, an antibody against the adhesion receptor E-selectin mediated liposome delivery to the HUVEC cell line.[24]

Finally, the majority of studies described above use chemical crosslinking to tether the targeting molecule to the vector surface. However, a potentially more practical alternative is to genetically link the ligand to a DNA-binding polypeptide through the construction and expression of chimeric fusion proteins with multiple activities. This approach was first implemented by Fominaya and Wels through the generation of a three-domain protein with a single chain antibody directed against the ErbB-2 receptor, a *Pseudomonas* endotoxin A domain to facilitate endosomal escape of the vector, and a GAL4 DNA-binding domain. When mixed with polylysine and complexed with DNA, this multifunctional protein mediated gene delivery to a human breast carcinoma cell line in vitro.[25] They have since constructed a more effective chimeric protein using a diptheria toxin endosomal escape domain.[26] The concept of a multifunctional molecule has since also been utilized with the generation of several polypeptides containing both integrin-binding RGD sequences and DNA-binding cationic residues for use in lipoplex and polyplex delivery.[6,27,28]

Application of Biological Diversity for Targeting Gene Delivery Vectors

Targeting gene delivery has most often been implemented with a ligand that nature has provided to bind to a cell surface antigen of interest. If there is no readily available ligand, as discussed above, one can create a targeting moiety for a given receptor by generating an antibody. However, raising new antibodies can be a time and labor intensive process that is generally most effective when one has already chosen a receptor to target. Furthermore, it would be highly advantageous to develop a method to efficiently generate ligands to target a particular cell when one may not even know what cell surface proteins are expressed by that cell type.

For over a decade, phage display technology, the presentation of large libraries of peptides or proteins on the surface of phage particles, has been exploited to screen and identify biological molecules for a variety of properties, particularly binding affinity.[29] Its ability to search through up to 10^{10} "solutions" to a given affinity problem is a powerful property for many applications since a complete, mechanistic understanding of the protein–protein interactions at the molecular level is not required and is often not available for complex biological systems.[30] As a result of these features, the application of phage display to the drug and gene delivery field has recently begun to enjoy success.[31–33]

In general, phage display involves the presentation of a protein library on a bacteriophage coat protein. It was found that certain coat proteins of phages, such as pIII of M13, are amenable to the genetic insertion of foreign sequences without disrupting overall phage functions. A library of random or nearly random oligonucleotides is inserted into the gene encoding the coat protein, resulting in the display of random peptides on a region of the virion surface that is sterically accessible. Although systems were initially developed for peptide display, methods were soon developed to engineer the phage coat proteins for the insertion of larger proteins such as libraries of antibodies.[34] A resulting peptide or antibody library is screened for bioactivities such as protein binding or targeting using a variety of methods depending on the specific application. Within the drug delivery field, recent work had attempted phage display selection against known specific cell markers, cells in culture, or even organ tissues in vivo.

Early work in developing novel peptide-targeting molecules focused on integrins, receptors of particularly great interest since specific integrins are overexpressed on many tumor cells.[35] Ruoslahti and co-workers isolated peptides that possess an RGD motif that selectively binds to certain classes of integrin receptors by "panning" phage-displaying peptides on purified, immobilized integrin receptors.[36,37] Later, they reversed this procedure by immobilizing the resulting peptide ligands and selecting for new peptides that mimicked the functional binding sites on the integrin receptors.[38] Ruoslahti and co-workers have since demonstrated that the RGD-containing peptide was successful for targeting the phage to tumors when injected intravenously into tumor-bearing mice.[39]

In some cases, if the cell marker proves difficult to purify, or if the goal is to target a particular cell type with no detailed prior knowledge of its cell surface receptor repertoire, it may be more desirable to pan the phage library against cells instead of purified receptors. Work has been conducted to isolate peptides that have affinities for fibroblasts[40] and vascular endothelial cells.[41] While these in vitro selection systems facilitate the isolation of peptides that bind cell surfaces, they lack efficient means for negative selection. That is, ligands that show high affinity binding to a cell could be binding to a common surface antigen expressed by many or even all cell types. Once a collection of phage that bind a cell of interest has been isolated, this problem of ligand promiscuity or low selectivity can be addressed by passing the library over nontargeted cells for ligand "subtraction." An alternative is to perform the selection in vivo, a type of screen particularly useful for organ or tissue targeting.

Pasqualini et al. conducted the first in vivo screen.[31] A phage display peptide library was injected intravascularly into mice for short time periods, and vasculature from individual organs, such as brain and kidney, was subsequently excised and incubated with *E. coli* for phage rescue. After several rounds of selection, peptides were isolated and showed selectivity for brain or kidney vasculature. Since this pioneering work, similar techniques have been used to isolate unique peptides that target vascular endothelium of various organs including lung, skin, and pancreas,[42] as well as tumor cells.[43] More recently, Ruoslahti and co-worker have also identified the receptors that direct the organ homing of these peptides,[44,45] work that may also aid our understanding of tumor biology. In addition, while a major drawback of the otherwise powerful in vivo selection approach is that it is limited to animal models, identifying cell and tumor specific proteins in these animals may lead to new specific targets for human therapeutic use.

In addition to phage, several other platforms have been developed for peptide and antibody library display, including bacterial cells[46] and yeast.[47] Brown et al. utilized an *E. coli* peptide display library to select for peptide sequences that bind tumor endothelial cell markers, and recurring peptide sequences were identified.[48] In addition, Nakajima et al. developed a bacterial display library of peptides fused to invasin, a bacterial surface protein.[49] Peptides that showed affinity for human VA13 cells were isolated upon screening the resultant fusion peptide library in vitro. Furthermore, although most cell targeting work has utilized peptides, recent studies have demonstrated the potential for targeting with antibodies. Marks and co-workers developed a single-chain antibody phage library and selected for targeting of breast tumor cells. Antibodies were identified to bind receptors on tumor cells, but not normal human cells, in vitro.[50,51]

In addition to directing the binding of vectors to specific cell surface receptors, it is also of great interest to select for ligands with the added ability to mediate cell entry. These are ligands that presumably bind surface antigens that undergo rapid internalization and turnover. One study by Marks and co-workers screened an antibody phage display library for the ability to bind and undergo endocytosis by a breast tumor cell line,[51] and other work characterized the parameters in the phage display of an antibody that optimized internalization.[50] This type of selection has been extended to the identification of ligands that mediate transcytosis to offer the potential of facilitating vector delivery through vascular endothelial cell layers to access the surrounding tissue. In one experiment, Ivanenkov et al. identified a peptide sequence that contained the putative RGD integrin receptor binding motif but also mediated basal to apical transcytosis of endothelial Madin–Darby canine kidney cells.[52] This type of system could be useful for developing vectors that are capable of overcoming vascular cell layers, such as the blood–brain barrier that obstructs drug delivery into the central nervous system. Finally, the development of phage that can mediate low efficiency gene delivery to mammalian cells may aid in the identification of peptides or proteins that facilitate vector passage through other steps of the gene delivery pathway.[53]

Exploring biological diversity to develop targeted vector systems is not limited to identifying novel ligands, i.e. targeting by vector surface engineering. Tissue specific promoters are promising alternatives for expression targeting,[54] but they too suffer from the difficulty of rationally designing molecules, in this case promoters, capable of cell targeting. In one novel solution to this problem, Li et al. generated a recombinant promoter library by randomly assembling myogenic regulatory or transcription factor binding elements.[55] Individual clones were screened for transcriptional activities in muscle cells both in vitro and in vivo. A number of the artificial promoters were found to possess transcriptional activity even greater than that of natural myogenic promoters. These promoters also proved to be specific for muscle cells, as expression in other cell types was significantly lower both in vitro and in vivo. Other tissue specific and tumor selective promoters have been identified, including ones for liver, neurons, breast cancer cells, and solid tumors.[56] Therefore, the combination of delivery and expressional targeting may eventually lead to the development of vectors capable of pinpointed gene expression.

Engineering Vectors for Stealth

Chemical Approaches for Synthesizing Protein Resistant Surfaces

A significant amount of work has been devoted to the identification and implementation of targeting proteins and peptides for specific gene delivery; however, another complementary property is also required of vector surfaces: stealth. It has been known for over a decade that nanoparticles, such as gene delivery vectors, are rapidly cleared from circulation by the reticulo-endothelial system, or macrophages of the liver and spleen.[57,58] In addition, protein adsorption to the vector surface, or opsonization, may lead to vector disabling or aggregation, followed by elimination from circulation by the innate immune system.[59–61] Therefore, for a vector particle to ever successfully reach a target cell after injection in vivo, it must first evade neutralization by serum components and elimination by cells of the immune system.

Fundamental work conducted in the past decade by the groups of Andrade, de Gennes, and Whitesides has identified and characterized polyethylene glycol (PEG) as a molecule that resists the adsorption of proteins.[62–65] The protein resistant properties of PEG have been attributed to a number of factors. First, PEG chains are conformationally very mobile and may sterically hinder the interactions of proteins with a coated surface. In addition, PEG in aqueous solution is a highly hydrated molecule, and the water structure near a coated surface may play a role in excluding protein binding. Finally, the protein-repelling property of PEG and analogous materials may be related to the fact that they contain polar functional groups, but no net charge or hydrogen bond donating groups. This fundamental work has since been applied to a number of applications, including drug and gene delivery as well as tissue engineering.

Grafting PEG onto the surface of synthetic vectors has improved their stealth in a number of studies. For example, early drug delivery studies found that PEG derivitization of liposomes can increase circulatory half-lives over tenfold while maintaining the potential for antibody-mediated targeting.[66,67] PEGylation of vehicles has subsequently been used to enhance targeted lipoplex delivery of genes or nucleic acids and has also been found to improve formulation stability.[68,69] Finally, crosslinking PEG to polyethylenimine was also found to reduce the interaction with blood proteins and increase the circulatory lifetime of polyplexes targeted to the transferrin receptor.[61]

In addition to PEG and other synthetic molecules that repel proteins, other more natural molecules may confer stealth on a particle. For example, the human immunodeficiency virus has learned that coating its envelope glycoprotein with over 50% carbohydrate by weight helps it evade immune detection. One study found that a clone of the HIV envelope that lacked a glycosylation consensus sequence at one site rapidly mutated under immune pressure to regain the oligosaccharide.[70] By analogy, oligosaccharides may offer synthetic systems the potential for immune evasion. Jaulin et al. found that coating nanoparticles with heparin or dextran reduced macrophage uptake in vitro.[71] The use of a modified cyclodextran as the polymeric material for gene delivery by Gonzalez et al. may also offer the same advantageous property.[72]

Biological Approaches for Enhancing Vector Stealth

PEGylation protects vectors from binding to serum components and subsequent inactivation by the innate immune system. As discussed above for HIV, pathogens in nature have similarly developed strategies for evading the immune system in blood stream.[73] This suggests that there may be classes of unexplored biological molecules (peptides, carbohydrates, or lipids) with the potential to resist opsonization of vectors.

Sokoloff et al. demonstrated another application of phage display: the identification of peptide sequences that promote the longevity of foreign particles in serum.[74] Phage coat protein was engineered to display random peptide sequences, yielding a library of phage particles that simulate delivery vehicles coated with peptides. Upon intravenous injection into rats or incubation in serum in vitro, complement and antibodies rapidly eliminated the majority of the library. However, examination of the minority of particles that survived in circulation led to the identification of consensus peptide sequences. In rat serum, peptides with a C-terminal lysine and arginine protects from inactivation by binding C-reactive protein. In contrast, particle stability in human serum was observed with tyrosine- containing peptides. These peptides were found to bind serum α-macroglobulin, which subsequently shielded the particles. This novel solution to the stealthing problem, camouflaging with endogenous blood components to prevent opsonization, differs from conventional "cloaking" strategies such as PEGylation of vectors. The differing results between rat and human serum additionally highlighted the importance of conducting human studies.

It remains to be seen whether these peptides would retain their activity when transplanted into a different context, such as the surface of a lipoplex or polyplex vector. However, this landmark work by Sokoloff et al. represents a promising alternative approach for the identification of materials for stealthing vectors.

Summary

Synthetic gene delivery vehicles have made significant advances in the past 15 years of research, particularly in the area of surface engineering. A number of natural ligands have been utilized for vector attachment and targeting to specific cells in vitro and in vivo. In addition, when one does not know the natural ligand for a particular surface protein, or even an appropriate surface receptor target expressed by a cell, the use of phage or bacterial display still offers the opportunity to identify ligands for targeted delivery. This process of peptide or antibody display for the identification of novel ligands conceptually resembles the natural evolutionary process that led viruses to bind to specific receptors. Furthermore, progress has been made in the development of chemical methods to create vectors that resist nonspecific protein or cellular binding for immune evasion. There are indications that these efforts to stealth synthetic vectors may also learn from biological approaches. Although they offer a number of potential advantages, synthetic vectors still require further progress in a number of areas, including surface engineering, before they can match the overall performance and efficiency of their viral counterparts. However, the continued application of complementary chemical and biological methods, guided by the imitation of nature, promises to narrow or even eliminate the gap.

Literature Cited

1. Mounkes LC, Zhong W, Cipres-Palacin G, Heath TD, Debs RJ. J Biol Chem 1998; 273:26164-26170.
2. Schaffer DV, Lauffenburger DA. Curr Opin Mol Ther 2000; 2:155-161.
3. Wu GY, Wu CH. J Biol Chem 1987; 262:4429-4432.
4. Wu GY, Wu CH. J Biol Chem 1988; 263:14621-14624.
5. Wang CY, Huang L. Proc Natl Acad Sci USA 1987; 84:7851-7855.
6. Hart SL, Collins L, Gustafsson K, Fabre JW. Gene Ther 1997; 4:1225-1230.
7. Erbacher P, Remy JS, Behr, JP. Gene Ther 1999; 6:138-145.
8. Colin M, Harbottle RP, Knight A, Kornprobst M, Cooper RG. Miller AD, Trugnan G, Capeau J, Coutelle C, Brahimi-Horn MC. Gene Ther 1998; 5:1488-1498.
9. Wickham TJ, Mathias P, Cheresh DA, Nemerow GR. Cell 1993; 73:309-319.
10. Gabius HJ. Eur J Biochem 1997; 243:543-576.
11. Ferkol T, Perales JC, Mularo F, Hanson RW. Proc Natl Acad Sci USA 1996 93:101-105.
12. Ferkol T, Mularo F, Hillard J, Lodish S, Perales JC, Ziady A, Konstan M, Am J Respir Cell Mol Biol 1998; 18:591-601.
13. Diebold SS, Kursa M, Wagner E, Cotten M, Zenke M. J Biol Chem 1999; 274:19087-19094.
14. Diebold SS, Lehrmann H, Kursa M, Wagner E, Cotten M, Zenke M. Hum Gene Ther 1999; 10:775-786.
15. Wagner E, Zenke M, Cotten M, Beug H, Birnstiel ML. Proc Natl Acad Sci USA 1990; 87:3410-3414.
16. Sudimack J, Lee RJ. Adv Drug Deliv Rev 2000; 41:147-162.
17. Hoganson DK, Chandler LA, Fleurbaaij GA, Ying W, Black ME, Doukas J, Pierce GF, Baird A, Sosnowski BA. Hum Gene Ther 1998; 9:2565-2575.
18. Sosnowski BA, Gonzalez AM, Chandler LA, Buechler YJ, Pierce GF, Baird A. J Biol Chem 1996; 271:33647-33653.
19. Schaffer DV, Lauffenburger DA. J Biol Chem 1998; 273:28004-28009.
20. Schaffer DV, Fidelman N, Dan N, Lauffenburger DA. Biotech & Bioeng in press.
21. Martinez-Fong D, Navarro-Quiroga I, Ochoa I, Alvarez-Maya I, Meraz MA, Luna J, Arias-Montano JA. Brain Res Mol Brain Res 1999; 69:249-262.
22. Mohr L, Schauer JI, Boutin RH, Moradpour D, Wands JR. Hepatology 1999; 29:82-89.
23. Li S, Tan Y, Viroonchatapan E, Pitt BR, Huang L. Am J Physiol Lung Cell Mol Physiol 2000; 278:L504-L511.
24. Spragg DD, Alford DR, Greferath R, Larsen CE, Lee KD, Gurtner GC, Cybulsky MI, Tosi PF, Nicolau C, Gimbrone MA. Proc Natl Acad Sci USA 1997; 94:8795-8800.
25. Fominaya J, Wels W. J Biol Chem 1996; 271:10560-10568.
26. Fominaya J, Uherek C, Wels W. Gene Ther 1998; 5:521-530.
27. Li JM, Collins L, Zhang X, Gustafsson K, Fabre JW. Transplantation 2000; 70:1616-1624.
28. Aris A, Feliu JX, Knight A, Coutelle C, Villverde A. Biotechnol Bioeng 2000; 68:689-696.
29. Smith GP. Science 1985; 228:1315-1317.
30. Daniels DA, Lane DP. Methods 1996; 9:494-507.
31. Pasqualini R, Ruoslahti E. Nature 1996; 380:364-366.
32. Koivunen E, Arap W, Rajotte D, Lahdenranta J, Pasqualini R. J Nucl Med 1999; 40:883-888.
33. Brown KC. Curr Opin Chem Biol 2000; 4:16-21.
34. Winter G, Griffiths AD, Hawkins RE, Hoogenboom HR. Annu Rev Immunol 1994; 12:433-455.
35. Ruoslahti E. Annu Rev Cell Dev Biol 1996; 12:697-715.
36. Koivunen E, Gay DA, Ruoslahti E. J Biol Chem 1993; 268:20205-20210.
37. Koivunen E, Wang B, Ruoslahti E. J Cell Biol 1994; 124:373-380.
38. Pasqualini R, Koivunen E, Ruoslahti E. J Cell Biol 1995; 130:1189-1196.
39. Pasqualini R, Koivunen E, Ruoslahti E. Nat Biotechnol 1997; 15:542-546.
40. Barry MA, Dower WJ, Johnston SA. Nat Med 1996; 2:299-305.
41. Nicklin SA, White SJ, Watkins SJ, Hawkins RE, Baker AH. Circulation 2000; 102:231-237.
42. Rajotte D, Arap W, Hagedorn M, Koivunen E, Pasqualini R, Ruoslahti E. J Clin Invest 1998; 102:430-437.
43. Ruoslahti E, Rajotte D. Annu Rev Immunol 2000; 18:813-827.
44. Rajotte D, Ruoslahti E. J Biol Chem 1999; 274:11593-11598.
45. Fukuda MN, Ohyama C, Lowitz K, Matsuo O, Pasqualini R, Ruoslahti E, Fukuda M. Cancer Res 2000; 60:450-456.

46. Georgiou G, Poetschke HL, Stathopoulos C, Francisco JA. Trends Biotechnol 1993; 11:6-10.
47. Boder ET, Wittrup KD. Nat Biotechnol 1997; 15:553-557.
48. Brown CK, Modzelewski RA, Johnson CS, Wong MK. Ann Surg Oncol 2000; 7:743-749.
49. Nakajima H, Shimbara N, Shimonishi Y, Mimori T, Niwa S, Saya H. Gene 2000; 260:121-131.
50. Becerril B, Poul MA, Marks JD. Biochem Biophys Res Commun 1999; 255:386-393.
51. Poul MA, Becerril B, Nielsen UB, Morisson, P, Marks JD. J Mol Biol 2000; 301:1149-1161.
52. Ivanenkov VV, Menon AG. Biochem Biophys Res Commun 2000; 276:251-257.
53. Kassner PD, Burg MA, Baird A, Larocca D. Biochem Biophys Res Commun 1999; 264:921-928.
54. Miller N, Whelan J. Hum Gene Ther 1997; 8:803-815.
55. Li X, Eastman EM, Schwartz RJ, Draghia-Akli R. Nat Biotechnol 1999; 17:241-245.
56. Nettelbeck DM, Jerome V, Muller R. Trends Genet 2000; 16:174-181.
57. Rolland A, Collet B, Le Verge R, Toujas L. J Pharm Sci 1989; 78:481-484.
58. Gaur U, Sahoo SK, De TK, Ghosh PC, Maitra A, Ghosh PK. Int J Pharm 2000; 202:1-10.
59. Zelphati O, Uyechi LS, Barron LG, Szoka FC, Jr. Biochim Biophys Acta 1998; 1390:119-133.
60. Dash PR, Read ML, Barrett LB, Wolfert MA, Seymour LW. Gene Ther 1999; 6:643-650.
61. Ogris M, Brunner S, Schuller S, Kircheis R, Wagner E. Gene Ther 1999; 6:595-605.
62. Jeon SI, Andrade JD. J Colloid Interf Sci 1991; 142:159-166.
63. Jeon SI, Lee JH, Andrade JD, De Gennes, PG. J Colloid Interf Sci 1991; 142:149-158.
64. Harris JM, Zalipsky S. Poly(ethylene glycol). Chemistry and Biological Applications, American Chemical Society, Washington, DC, 1997.
65. Chapman RG, Ostuni E, Takayama S, Holmlin RE, Yan L, Whitesides GM. J Am Chem Soc 2000; 122:8303-8304.
66. Klibanov AL, Maruyama K, Torchilin VP, Huang L. FEBS Lett 1990; 268:235-237.
67. Mori A, Klibanov AL, Torchilin VP, Huang L. FEBS Lett 1991; 284:263-266.
68. Wang S, Lee RJ, Cauchon G, Gorenstein DG, Low PS. Proc Natl Acad Sci USA 92:3318-3322.
69. Hong K, Zheng W, Baker A, Papahadjopoulos D. FEBS Lett 1997; 400:233-237.
70. Schonning K, Jansson B, Olofsson S, Hansen JE. J Gen Virol 1996; 77:753-758.
71. Jaulin N, Appel M, Passirani C, Barratt G, Labarre D. J Drug Target 2000; 8:165-172.
72. Gonzalez H, Hwang SJ, Davis ME. Bioconjug Chem 1999; 10:1068-1074.
73. Wiertz EJ, Mukherjee S, Ploegh HL. Mol Med Today 1997; 3:116-123.
74. Sokoloff AV, Bock I, Zhang G, Sebestyen MG, Wolff JA. Mol Ther 2000; 2:131-139.

CHAPTER 3

Histidine Containing Peptides and Polypeptides as Nucleic Acid Vectors

Patrick Midoux, Eric LeCam, Dominique Coulaud, Etienne Delain, and Chantal Pichon

Abstract

Nucleic acid transfer in mammalian cells is drastically improved with devices which increase their delivery in the cytosol upon endocytosis. In this chapter, we describe the effect on plasmid DNA (pDNA) and oligonucleotide (ODN) transfer, of an histidine-rich peptide (H5WYG), histidylated oligolysine (HoK), and histidylated polylysine (HpK) designed on the basis of the membrane destabilization capacity of poly-L-histidine at a pH close to that of the endosomes. We report that H5WYG, which permeabilizes the cell membrane at pH 6.4, favors the transfection mediated by lactosylated polylysine/pDNA complexes and, by lowering the pH of extracellular medium, allows the loading of the cytosol and the cell nucleus with ODN. We show that HoK forms small cationic spherical particles of 35 nm with ODN and HpK rod or toroid cationic particles of 100 nm with pDNA. PEGylation stabilizes these particles at physiological salt concentration. We also show that (i) HoK/ODN complexes yield a more than 20-fold increase of the biological activity of antisense ODN towards the inhibition of transient as well as constitutive gene expression and (ii) HpK/pDNA complexes yield a transfection efficiency of 3–4.5 order of magnitude higher than do polylysine/pDNA complexes. We also provide evidence that the effect of these polyhistidylated molecules is mediated by imidazole protonation in endosomes. Overall our data show that polyhistidylated molecules constitute interesting devices for an efficient cytosolic delivery of nucleic acids, and that ionic complexes between histidylated polylysine and a pDNA are attractive for developing a nonviral gene delivery system.

Abbreviations: AcH, *N*-acetylated histidyl residue; BET, ethydium bromide; BOP, benzotriazol-1-yl-oxy-tris-(dimethylamino) phosphonium hexafluorophosphate; DIEA, diisopropylethylamine; DMSO, dimethylsulfoxide; DP, degree of polymerization; *Flu*-, fluorescein-labeled; His, histidyl residue; HoK, histidylated oligolysine; HpK, histidylated polylysine; ICAM-1, intercellular adhesion molecule-1; Lac, lactosyl residue; ODN, oligonucleotides; Osu, *N*-hydroxysuccinimidyl residue; pDNA, plasmid DNA; PEG, polyethyleneglycol; PEI, polyethyleneimine; pLK, poly-L-lysine; PS-ODN, phosphorothioate oligonucleotides; QELS, Quasi Elastic Light Scattering; TEM, Transmission Electron Microscopy; TFA, trifluoroacetic acid; TNF-α, tumor necrosis factor-α.

Synthetic DNA Delivery Systems, edited by Dan Luo and W. Mark Saltzman.

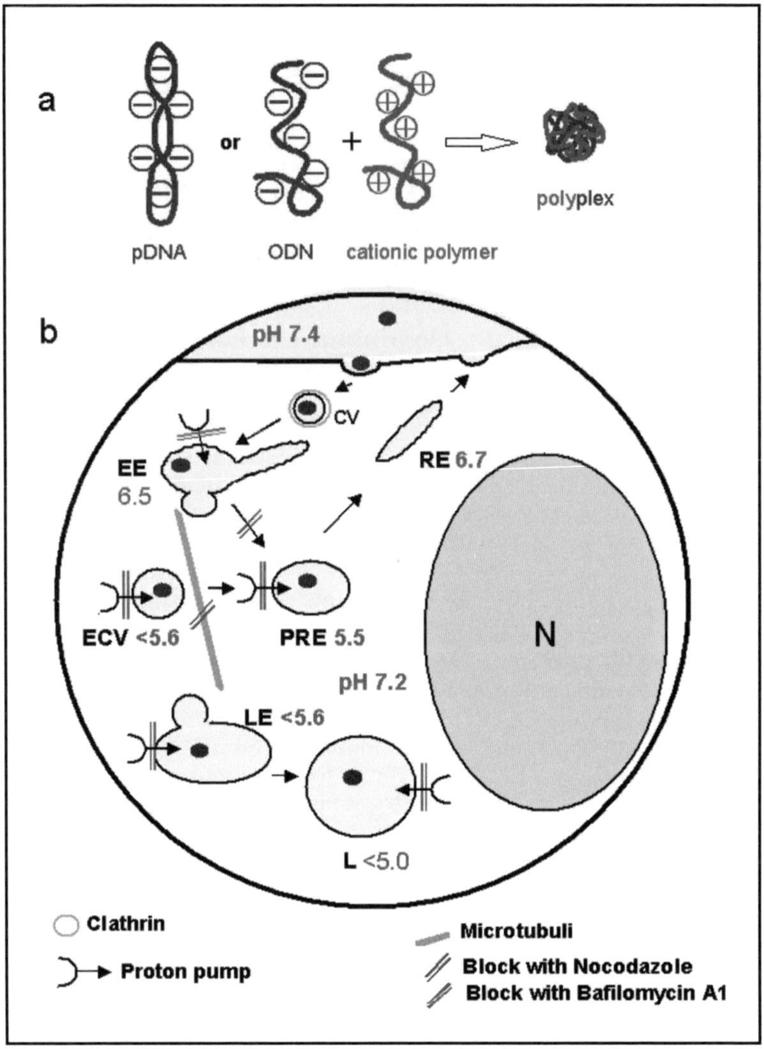

Fig. 3.1. (a) Schematic pDNA and ODN polyplexes; (b) endocytosis pathway and intracellular pH. CV = clathrin vesicle; EE = early endosome; RE = recycling endosome; ECV = endosomal carrier vesicle; PRE = perinuclear recycling endosome; LE = late endosome; L = lysosome; N = nucleus.

Introduction

The cellular delivery of nucleic acids with an expectation to get therapeutic benefits is not efficient because of their physicochemical features and requires vectors, preferably nonviral vectors to achieve it. Over the last decade, several cationic polymers were designed as vectors to transfect mammalian cells.[1–3] The self-assembly of plasmid DNA (pDNA) with cationic polymer induces pDNA condensation and the formation of polyplexes (DNA/cationic polymer complexes) which are efficiently taken up by cells (Fig. 3.1a). However, the intracellular trafficking dynamics of polyplexes and the mechanism allowing pDNA to reach the cell nucleus are still unknown. Microscopy analyses have shown that most polyplexes route via acidic vesicular compartments (endosomes or phagosomes), which are not lysosomes because they are not labeled with antilysosomal protein-specific antibodies.[4] No pDNA has been detected in the cytosol, and its location in the cell nucleus is a rare event. Therefore, the membrane of these vesicles constitutes one major barrier for the pDNA delivery in the cytosol before reaching the cell nucleus. On the basis of the knowledge of the endocytic pathway (Fig. 3.1b), strategies using weak bases such as chloroquine[5–9] or fusogenic/permeabilizing peptides[8,10–16] suitable to destabilize and/or disrupt acidic vesicles have been developed to improve the transmembrane passage through endocytic vesicles and thereby the nucleic acids delivery in the cytosol.

Antisense oligonucleotides (ODN) are also attractive drugs to inhibit or regulate gene expression.[17] As with pDNA, the self-assembly of ODN with cationic polymers such as polylysine conjugates.[18–22] and polyethyleneimine[23] has also been evaluated to increase ODN delivery. But despite the enhanced cell uptake, the increase of their biological activity required the presence of auxiliary factors to help them cross the endosomal membrane. Chloroquine was found to enhance significantly the biological activity of ODN, but peptides derived from the N-terminal fragment of the HA2 subunit of the influenza virus hemagglutinin did not, even when they were linked to ODN.[24–26] Studies on the intracellular location of ODN have pointed out that in a majority of cell types, most of ODN are confined inside vesicles and only a small amount finds its way to reach the RNA or DNA targets.[27–28] In contrast to pDNA, once in the cytosol, ODN penetrate rapidly into the nucleus because of their low molecular mass.[29–33] Therefore, an increase of the ODN delivery into the cytosol upon endocytosis is expected to increase their biological activity.

This chapter is focused on the generation of devices including a peptide and polypeptides by utilizing the imidazole group of histidine (pKa = 6.0) and the features of poly-L-histidine to destabilize endosomes in order to induce the cytosolic transfer of both ODN and pDNA (Fig. 3.2). We summarize their synthesis, their influence on pDNA and ODN transfer including the cellular uptake, the intracellular location, and the biological activity.

Peptide H5WYG

We have designed the peptide H5WYG (GLFHAIAHFIHGGWHGLIHGWYG), analogous to the N-terminal segment (GLFGAIAGFIEGGWTGMIDGWYG) of the HA2 subunit of the influenza virus hemagglutinin in which G-4, G-8, E-11, T-15, and D-19 were replaced by histidyl residues and M-17 by a leucyl residue.[34] Absorbance, fluorescence, and circular dichroism spectra indicated that the peptide underwent a dramatic conformational change between pH 7.0 and pH 6.0, which correlated with the protonation of histidines. At neutral pH, H5WYG aggregates and is weakly water-soluble but at pH 5.0 when all histidines are protonated, H5WYG is more soluble in water and exhibits unordered structures. This peptide permeabilizes cell membranes at pH 6.8 and thus is a suitable helper for delivering nucleic acids in the cytosol upon permeabilization of either endosomes after acidification of their lumen or the plasma membrane after acidification of the extracellular medium.

Effect of H5WYG on Polyfection

Polyfection with lactosylated polylysine (Lac-pLK) has been shown to be efficient on various cell types expressing a galactose-specific lectin on their cell surface. But, the efficiency is strictly dependent on the presence of chloroquine or fusogenic/permeabilizing peptides such as E5 peptides.[8,16] Human hepatocarcinoma cells (HepG2 cells), murine melanoma cells (B16 cells), and rabbit vascular smooth muscle cells (Rb-1 cells) have

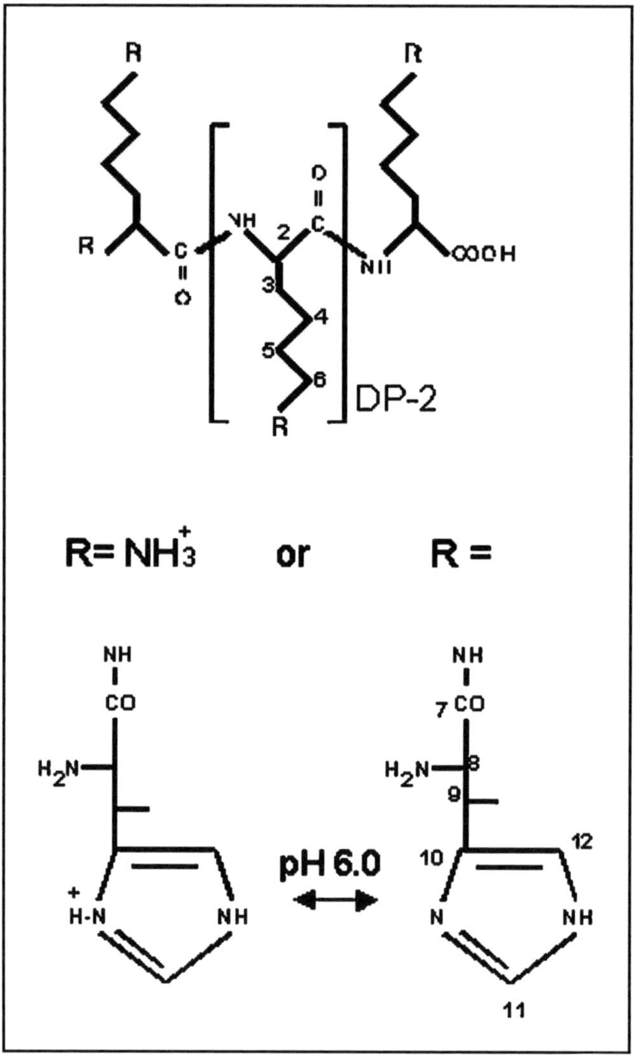

Fig. 3.2. Schematic structures of histidylated polylysine (HpK; DP = 190) or histidylated oligolysine (HoK; DP = 19).

served to test the ability of H5WYG to increase the polyfection efficiency. In Table 3.1, we show a comparison between transfections with a pDNA encoding luciferase reporter gene complexed with Lac-pLK in the absence or in the presence of either H5WYG or its anionic counterpart E5WYG. As expected, when the transfection is performed in the absence of any helper peptide, the luciferase activity is low. In the presence of H5WYG, the luciferase activities increase 93-, 215-, and 630-fold in HepG2, B16, and Rb-1 cells, respectively. The effect of H5WYG is not also impaired in the presence of serum (even at 50% serum concentration in the medium), in contrast to E5WYG which is almost inactive at few serum concentration.[34] Several limiting factors can explain the disparity between the efficiency of H5WYG and E5WYG on these different cell types. They include (i) the

Table 3.1. Effect of H5WYG on the polyfection with pCMVLuc/Lac-pLK complexes

	RLU × 10⁶		
Cells	**−Peptide**	**+H5WYG**	**+E5WYG**
B16	0.13	29	1.05
HepG2	0.1	9.3	52.5
Rb-1	0.1	63	1

Note. Cells were transfected for 4 h with 5 μg pDNA encoding luciferase complexed with 15 μg Lac-pLK in the absence or in the presence of 10 μM H5WYG or 10 μM E5WYG. RLU, the number of relative light units, represents the luciferase activity upon 48 h in 10^6 cells. From Ref. (34).

absorption of peptides to serum proteins of the extracellular medium, (ii) the uptake of both polyplexes and peptides, (iii) the intravesicular peptide and polyplexes concentrations, and (iv) the luminal pH of the compartment where they are located. It is obvious that an optimal permeabilization can be achieved in endosomes having an optimal luminal pH combined with the right peptide concentration. The fact that H5WYG is uncharged at neutral pH and does not ionically interact with serum proteins, combined with its ability to permeabilize membrane at pH 6.8 such as in early endocytic compartments, could explain the wide efficiency of H5WYG and make it a potential helper for pDNA cytosolic delivery.

Cytosolic and Nuclear Delivery of Oligonucleotides by H5WYG

Previously, we have reported that ODN uptake can be increased by complexing them with glycosylated polylysine.[21] However, their biological activity is strictly dependent on the presence of chloroquine. We have used H5WYG to favor ODN delivery in the cytosol. However, an incubation of cells at 37°C with both *Flu*-ODN and H5WYG did not allow ODN delivery in the cytosol because the peptide concentration inside endocytic vesicles was too weak to allow membrane destabilization. Nevertheless, we found that H5WYG could be used to easily load the cytosol and the cell nucleus with *Flu*-ODN by lowering the extracellular pH of medium containing both *Flu*-ODN and H5WYG (Fig. 3.3b and d). *Flu*-ODN does not enter in the cytosol in the presence of H5WYG at neutral pH (Fig. 3.3a) nor at pH 6.5 in the absence of H5WYG (Fig. 3.3c). The viability of the cells is preserved better with H5WYG than with anionic peptides such as E5WYG which require a more acidic pH to permeabilize membrane.[35] This simple procedure may be useful for introducing large amounts of ODN in the cytosol and in the nucleus to study the interactions of antisense ODN with their RNA targets, of decoys with their protein target, and of ODN-forming triple helices with their DNA target.

DNA Transfer by Using Histidylated Polylysines (HpK)

The substitution of poly-L-lysine with several histidyl residues was performed to combine the pDNA compaction capacity of polylysine and the membrane destabilization features of poly-L-histidine at a pH close to that of the endosome lumen to favor the pDNA cytosolic delivery.

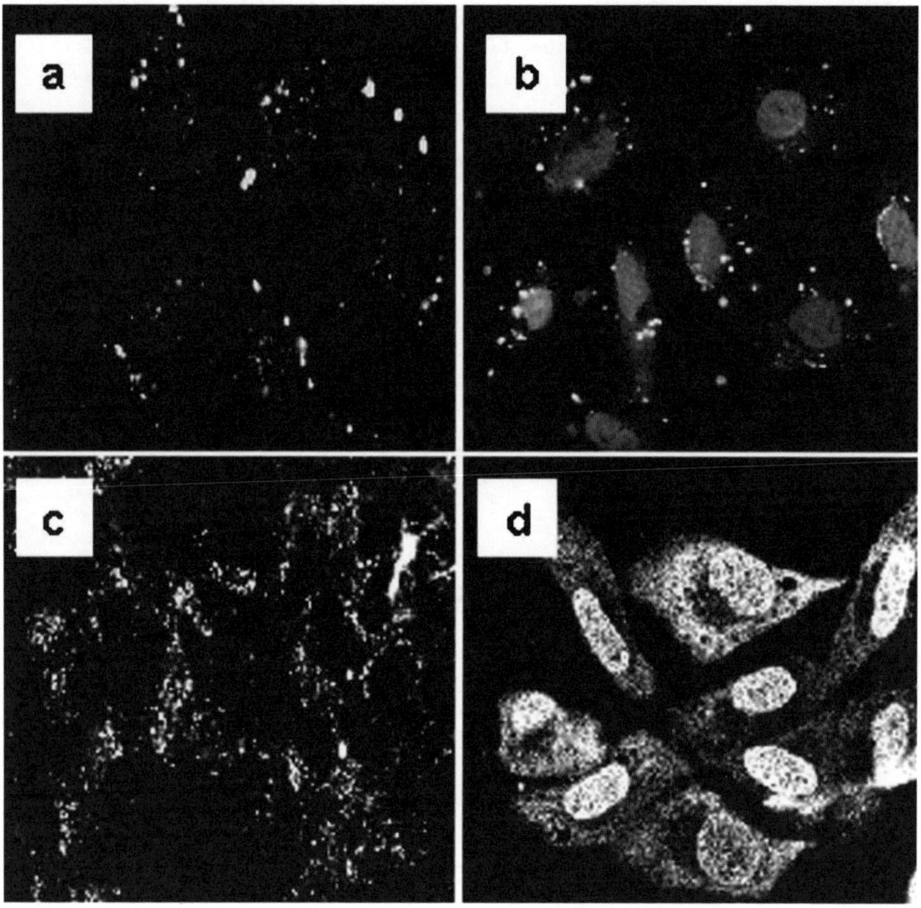

Fig. 3.3. H5WYG-mediated cytosolic delivery of ODN. A549 cells were incubated for 30 min at 20°C with 0.1 μM *Flu*-PS-ODN, and 20 μM H5WYG at either (a) pH 7.4 or (b) pH 6.5. Rb-1 cells were incubated for 30 min at 20°C with 0.1 μM *Flu*-PS-ODN at pH 6.5 (c) in the absence or (d) in the presence of 20 μM H5WYG. Then, the medium was removed and cells were further incubated for 30 min at 37°C at pH 7.4. Cells were fixed with 2% *p*-formaldehyde and visualized with a confocal microscope.

Synthesis of Histidylated Polylysine (HpK)

First, pLK (poly-L-lysine, HBr; average molecular weight of 30,000–50,000; DP = 190) is modified into pLK *p*-toluene sulfonate salt to remove cytotoxic bromide ions. PLK *p*-toluene sulfonate salt in DMSO with an excess of DIEA is reacted with either (Boc)His(Boc)-OH in the presence of BOP or BocHis(Boc)-Osu. N-protecting Boc groups are removed by acidic treatment with an H_2O/TFA mixture (1:1; v/v). Water and TFA are removed under reduced pressure. HpK is precipitated with isopropanol, collected by centrifugation, solubilized in distilled water, and freezed-dried. The average number of His bound per pLK molecule is determined by ^{1}H-NMR spectroscopy according to $x = 6 \cdot (h_{8.7}/h_{\mathrm{Lys}}) \cdot \mathrm{DP}$, where $h_{8.7}$ is the value of the integration of the signal at 8.7 ppm corresponding to the proton (C_{12}) of His (Fig. 3.2), h_{Lys} is that in the range from 1.3 to

1.9 ppm corresponding to the 6 methylene protons (C_3, C_4, and C_5) of lysine residues (Fig. 3.2), and DP is the degree of polymerization of the polymer.

Histidylated Polyplexes

The self-assembly of pDNA in an aqueous solution via cooperative electrostatic interactions with HpK induces pDNA condensation and the formation of polyplexes (Fig. 3.1a). The size and the homogeneity of polyplexes which depend on the procedure for their preparation can be assessed by Quasi Elastic Light Scattering (QELS) and Transmission Electron Microscopy (TEM). QELS measurements show that homogenous polyplexes with an average diameter of 130 nm are obtained when they are prepared in distilled water or in a neutral buffer with a low salt concentration (Fig. 3.4a). The global charge (ζ potential)

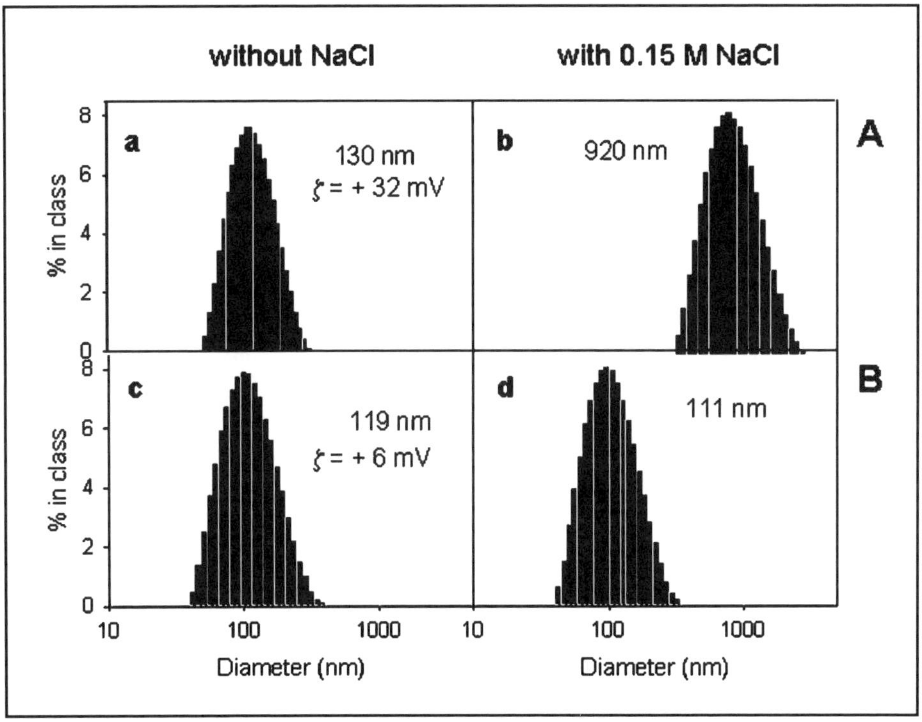

Fig. 3.4. Size and charge of His-polyplexes; effect of PEGylation. Twenty micrograms pDNA of 5 kb in 150 μL of 10 mM Hepes buffer, pH 8.0, were mixed with 60 μg HpK in 60 μL of the same buffer. (A) After 30 min at 20°C, 1 mL of either (a) distilled water or (b) PBS was added; (B) after 30 min at 20°C, 80 μg PEG-Osu (Mw 5000, Sigma) (PEG/HpK molar ratio of 18) in 10 μL DMSO was added. After 2 h at 20°C, 1 mL of either (c) distilled water or (d) PBS was added. QELS measurement was performed after 1 h by using a ZetaSizer 3000 (Malvern Instruments, Orsay, France) with a sample refractive index of 1.59; viscosity of 0.89. The measurements were performed 10 times, and the size calculation was performed by using the automatic mode. The system was calibrated by using the 200 ± 5 nm standard dts 5050 from Malvern. The ζ potential of polyplexes was measured at low salt concentration by using the ZetaSizer 3000 with the following parameters: viscosity, 0.891 cP; dielectric constant, 79; temperature, 25°C; F(Ka), 1.50 (Smoluchowsky); maximum voltage of the current, 15 V. The measurements were performed five times with zero field correction. The system was calibrated by using the −50 ± 5 mV standard dts 5050 from Malvern.

of polyplexes is +32 mV and +18 mV in distilled water and in 10 mM Hepes buffer, pH 7.4, respectively. Unfortunately, large aggregates of 920 nm in average are obtained at physiological salt concentration (Fig. 3.4b). But polyplexes are stabilized at 100 and 50 nm after diluted 10 times in 0.15 M NaCl buffer containing 10 and 50% bovine fetal serum, respectively.[36] This stabilization is due to the absorption of several anionic serum proteins including albumin on polyplexes, preventing hydrophobic contacts between them. In addition, the global charge of serum-stabilized polyplexes is reversed to −25 mV.[36] As a consequence, serum-stabilized polyplexes have a reduced nonspecific binding on the cell surface via the negatively charged sulfated proteoglycans. The aggregation of polyplexes at physiological salt concentration can also be inhibited by coupling polyethyleneglycol (PEG) either on polyplexes or on polymer before the formation of polyplexes.[37,38] Upon PEGylation, polyplexes exhibit an average size of 119 nm and a reduced ζ potential of +6 mV (Fig. 3.4c). The size remains at 111 nm after 1 h at physiological salt concentration (Fig. 3.4d). Moreover, PEGylated polyplexes remain stable for several days. Polyfection using PEGylated histidylated polyplexes is as efficient as serum-stabilized histidylated polyplexes. TEM observations show that polyplexes form both toroids and rods (Fig. 3.5). This low magnification picture shows clearly the diversity in shape and size of the condensed polyplexes. It is even possible to see DNA segments, not yet fully condensed, eventually attached to rods or to small dense dots. Given the molecular mass of a 5-kb pDNA and a HpK such as pLK substituted with 80 His, the molecular mass of a complex with a polymer/DNA of 1:3 (μg:μg) is about 13,300,000, corresponding to a volume of about 1.5×10^4 nm^3 and to a sphere with a diameter of about 31 nm. The average diameter of 130 nm obtained from QELS measurements is much higher than that of one single pDNA condensed particle. Indeed, TEM shows that their size is actually variable (Fig. 3.5). To determine how many DNA molecules can be compacted in a single toroid, linear DNA (Fig. 3.6c) was condensed with gluconoylated polylysine.[40] Intermediate stages (Fig. 3.6a and b) show that up to eight DNA ends are detectable (arrows, Fig. 3.6a), indicating that roughly a minimum of four DNA segments could be included in a single toroid or a spherical aggregate. A fully condensed toroid is shown in Fig. 3.6d, as a final step of condensation (same magnification as Fig. 3.6a and b).

Polyfection

Based on the luciferase activity, the polyfection of HepG2 cells with 10 μg of a pDNA encoding the luciferase gene, complexed with 30 μg of HpK, is 3–4.5 order of magnitude higher than polyfection with the pDNA complexed with pLK in the absence or in the presence of chloroquine or a fusogenic peptide.[41] The luciferase activity is maximal when cells are incubated for 3–4 h in the presence of HpK/pDNA complexes. No cytoxicity was observed. For HepG2 cells, optimal transfections were obtained with polyplexes made with HpK containing 63–81 His. Similar results were found for other cell lines. The luciferase activity in human embryonic kidney cells (293 cells) is maximal when polyplexes are made with HpK containing 50–100 His (Fig. 3.7). The effectiveness is dependent on the proportion between the size of pLK and the substitution level. The transfection efficiency is similar when polyplexes are made with a pLK of DP 190 substituted with 44% His and a pLK of DP 72 or DP 36 with about 20% His (Fig. 3.8). The efficiency decreases drastically when pLK of DP 72 and that of DP 36 are substituted with 43 and 53% His, respectively. Although polyplexes made with pLK substituted with 4-carboxymethyl-imidazolyl residues are about twofold more efficient than those made with HpK, the substitution of pLK with histidyl residues is preferred (Table 3.2). Indeed, 4-carboxymethyl-imidazole substitution

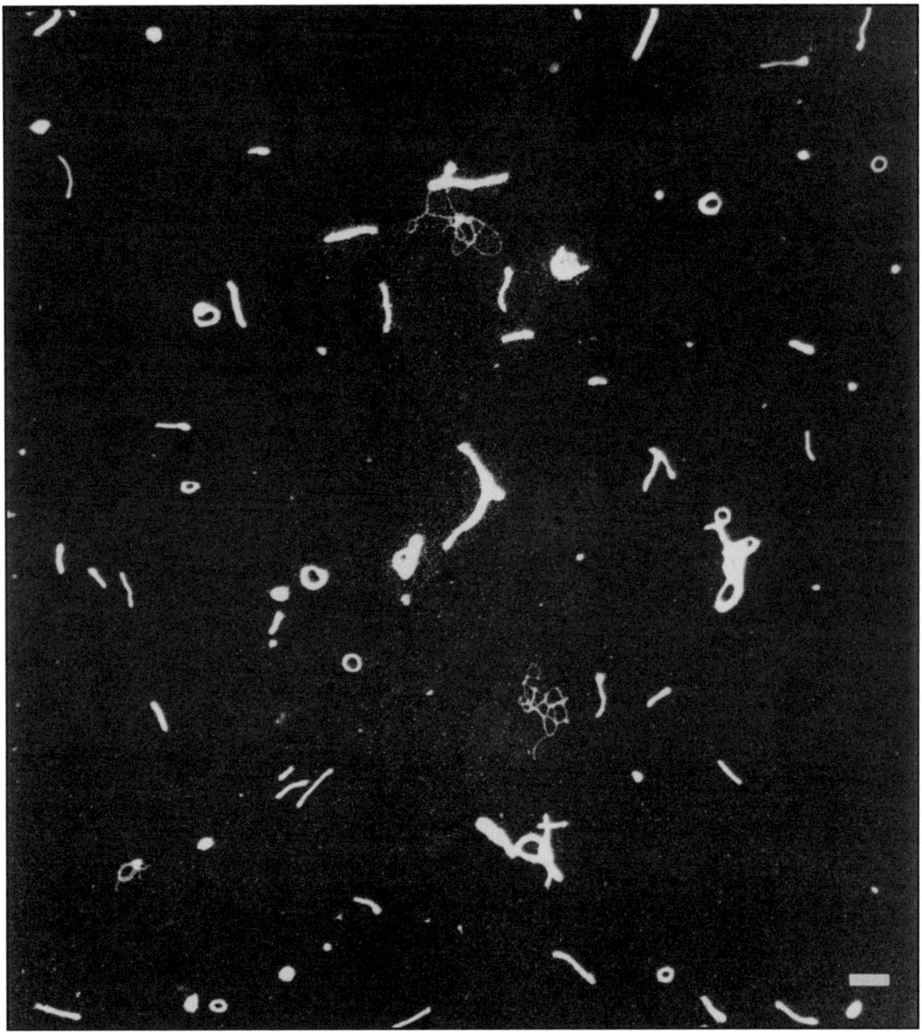

Fig. 3.5. Transmission Electron Microscopy of PEG-HpK/pDNA. Ten micrograms pDNA of 5 kb in 70 μL of 10 mM Hepes buffer, pH 7.4, were complexed with 30 μg of PEG-HpK in 30 μL of the same buffer. After 30 min at 20°C, 1 mL of 10 mM Hepes buffer, pH 7.4, was added. Five microliters of this sample was deposited on a carbon-coated grid activated by a glow-discharge in the presence of pentylamine, washed with 2% uranyl acetate, blotted and dried. Observation is done in the dark-field mode with a LEO-Zeiss 902 electron microscope, filtering out inelastic electrons with the electron spectrometer for better imaging.[39] Scale bar = 100 nm.

often gives polymers an uncontrolled brown color during the synthesis. Methyl substitution of the nitrogen atoms of the imidazole ring gives less efficient vectors, and acetylation of the α-amino group of His decreases slightly the transfection efficiency (Table 3.2). We have found that the presence of the α-amino group of His increases pDNA condensation and results in formation of smaller polyplexes. In addition, it elevates the pK of the imidazole protonation to pH 6.9, which is better to destabilize early endocytic vesicles.

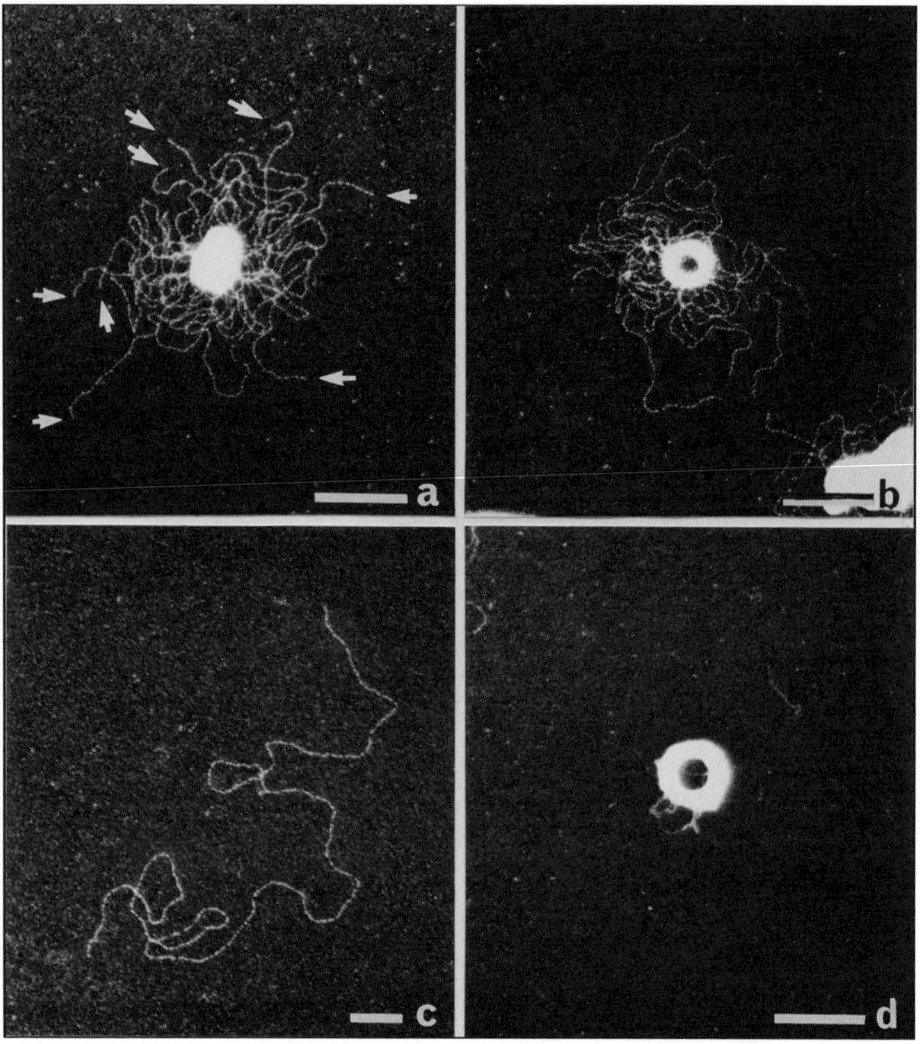

Fig. 3.6. Polyplexes formation. Ten micrograms of linear pDNA of 5 kb in 70 μL of 10 mM Hepes buffer, pH 7.4, was complexed with 5 μg of gluconoylated pLK in 30 μL of the same buffer. After 30 min at 20°C, 1 mL of 10 mM Hepes buffer, pH 7.4, was added, the sample was processed and observed as for Fig. 5. In the process of condensation, ends of the linear plasmids remain visible providing an indication that 4 or more DNA species are engaged in the formation of the dense aggregates (a) or toroids (b, d); (c) shows the control linear plasmid. Scale bars = 100 nm.

Intracellular Trafficking

The mechanisms of polyplexes entry and intracellular trafficking are not yet understood. Taking their size into account, both endocytosis of polyplexes <100 nm and phagocytosis of bigger polyplexes can be involved. TEM observations of cystic fibrosis human airway epithelial cells ($\sum$CFTE cells) show that HpK/pDNA aggregates of various sizes (arrows) bind onto the plasma membrane and are taken up via vesicles (Fig. 3.9a–c). Particles engulfed from an expansion of the plasma membrane, similar to phagocytosis,

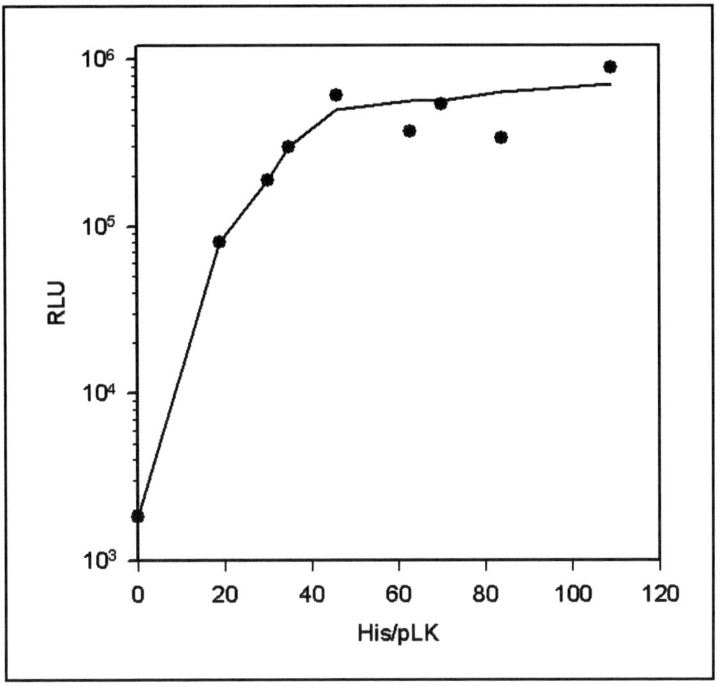

Fig. 3.7. Influence of the His/pLK ratio on HpK-mediated polyfection. Human embryonic kidney cells (293 cells; 3×10^5 cells plated in a 4 cm^2 well) were incubated for 4 h at 37°C in the presence of 10% FBS with 5 μg pCMVLuc complexed with 15 μg pLK (DP = 190) substituted with various His numbers. Then, cells were washed and cultured in complete culture medium containing 10% FBS. The gene expression was determined 48 h later by assaying the luciferase activity in cell lysates. RLU, the number of relative light units, represents the luciferase activity in 10^6 cells.

are observed for aggregates (Fig. 3.9a and b). Smaller particles (<200 nm) could be taken up via plasma membrane invagination as in endocytosis. The presence of vesicles containing several smaller particles (125–200 nm) seems to indicate that aggregates are dissociated soon after internalization (Fig. 3.9c, left arrow). Sometimes, small polyplexes (about 40 nm) are visible in the cell nucleus (Fig. 3.9d). Confocal microscopy analyses of vesicles containing fluorescent polyplexes have shown that they are not colocalized with compartments labeled with anti-LAMP1 antibodies, which is in favor for a nondelivery in a lysosomal compartment.[4] The intracellular trafficking of histidylated polyplexes requires more attention to clarify this peculiar routing.

ODN Transfer by Using Histidylated Oligolysines (HoK)

ODN/HoK Complexes

Histidylated oligolysines are prepared as described previously for HpK by substitution of oligolysine, *p*-toluene sulfate salt (DP 19) with His. The mixing in PBS of a 25 mer PS-ODN with HoK containing 15 His induces their self-assembly via electrostatic interactions (Fig. 3.1a) and can be followed by measuring the light scattering at 600 nm induced by complexes. Light scattering is maximal when 1 μM ODN is mixed with 2–4 μM HoK,

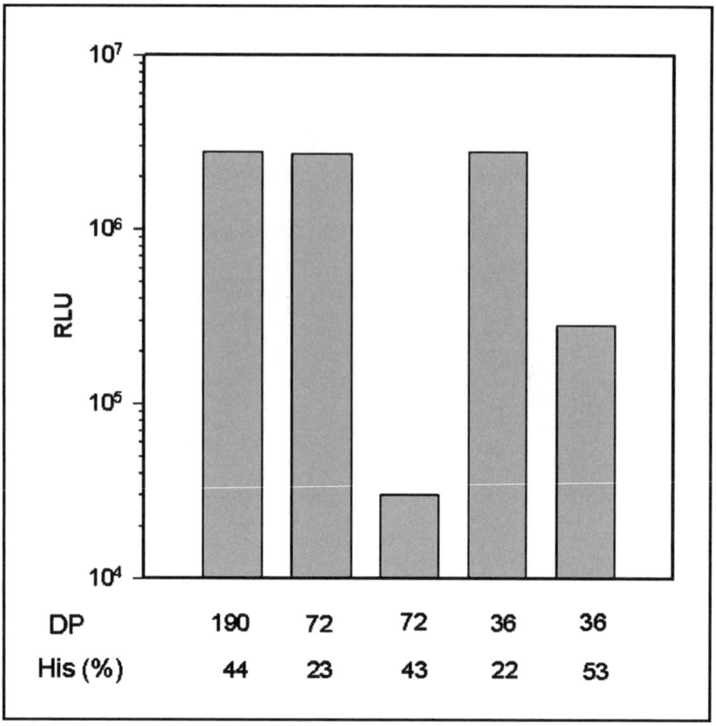

Fig. 3.8. Influence of the polylysine size on HpK-mediated polyfection. HepG2 cells (3×10^5 cells plated in a 4 cm^2 well) were incubated for 4 h at 37°C in the presence of 10% FBS with 5 μg pCMVLuc complexed with 15 μg pLK of various DP substituted with various His numbers. Cells were cultured and the gene expression was determined as described in Fig. 3.7.

Table 3.2. Influence of various histidyl residues on the polyfection efficiency

pLK substitution	**% of transfection**
histidyl	100
1-methyl-histidyl	57
3-methyl-histidyl	31
N-acetyl-histidyl	77
4-carboxymethyl-imidazolyl	172

Note. HepG2 cells were transfected with 5 μg pCMVLuc complexed with 15 μg pLK substituted with various types of histidyl groups. After 48 h, the luciferase activity was measured and expressed as the percentage relative to that obtained in cells transfected with HpK.

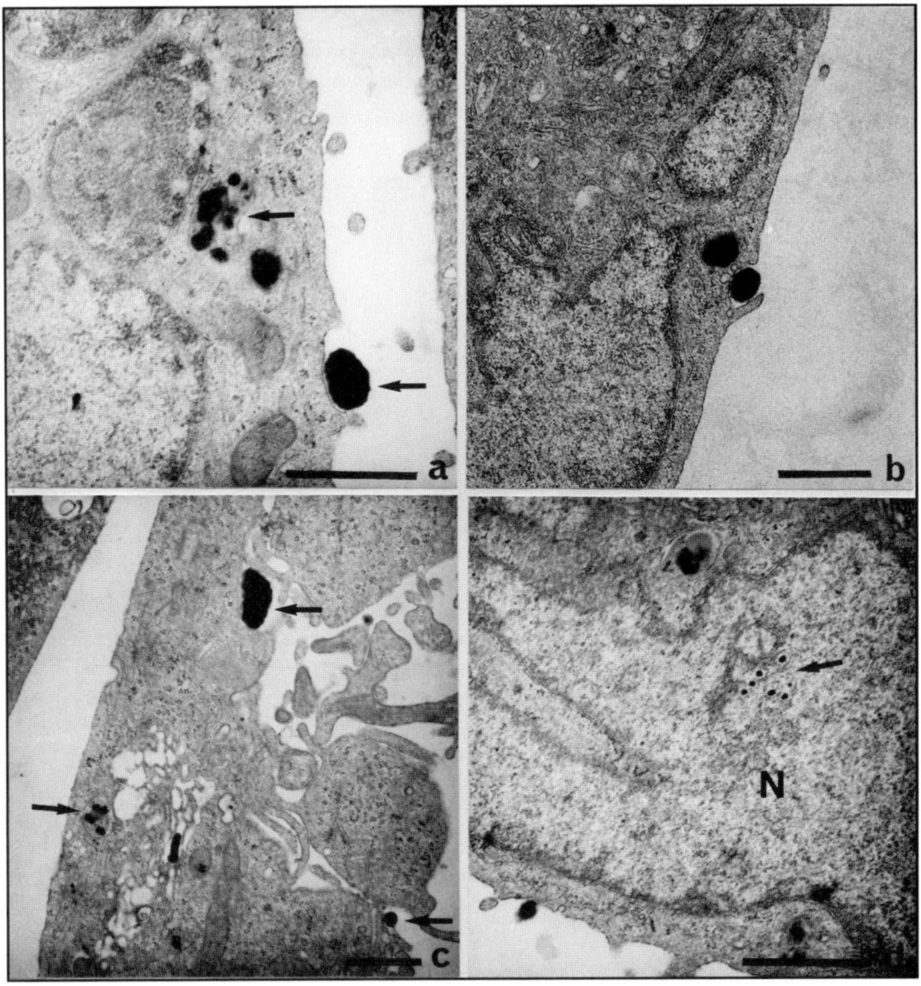

Fig. 3.9. Uptake of PEG-HpK/pDNA complexes. ∑CFTE cells incubated for 1 h at 37°C with HpK/pDNA complexes were washed and fixed for 30 min in 2.5% glutaraldehyde in PBS. Cells were harvested, dehydrated, and embedded. Stained sections were observed by traditional Transmission Electron Microscopy. Polyplexes of various sizes are visualized as dense "dots" in the electron micrographs (arrows). N = nucleus. Scale bar = 1 μm.

indicating that all ODN are complexed at a HoK/ODN molar ratio of 2:4. These complexes tend to aggregate in an aqueous solution at physiological salt concentration. As for pDNA/HpK complexes, aggregation of HoK/ODN complexes is prevented by PEGylation. QELS measurements show that PEG-HoK and ODN in 10 mM Hepes buffer, pH 7.4, form complexes with a size varying from 25 to 34 nm (Fig. 3.10). Complexes with a size of 35 nm remain stable in the presence of 0.15 M NaCl, even after 2 weeks at 4°C (Fig. 3.10). TEM observations show that these complexes are dense spherical particles with a homogenous size of 20–30 nm comparable to that found by QELS measurements (Fig. 3.11). These PEG-HoK/ODN complexes exhibit a shape and a size similar to that of complexes between PS-ODN (24 mer) and PEG-PEI (10 kDa molecular weight)[42]

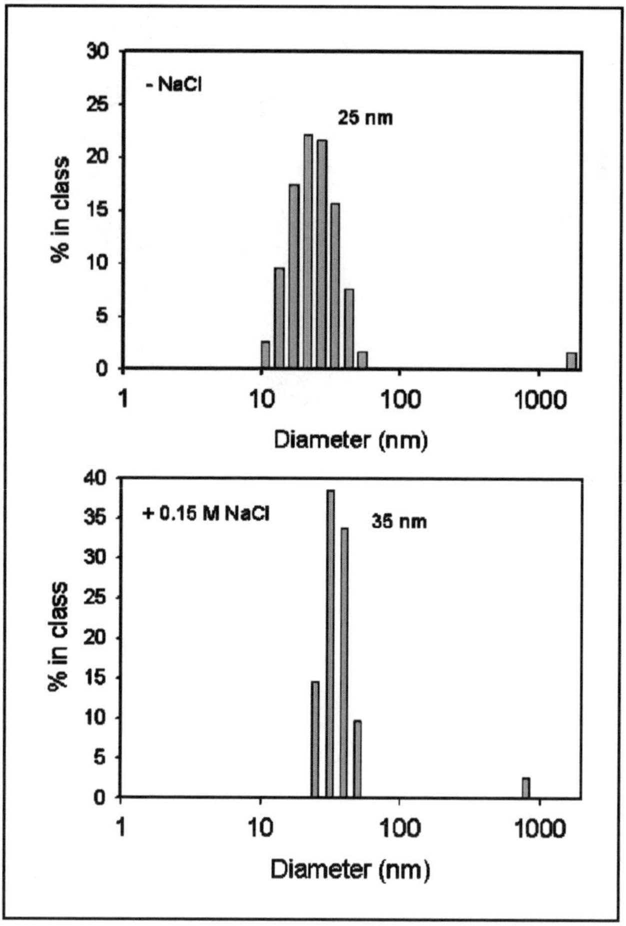

Fig. 3.10. Size of PEG-HoK/ODN complexes. One hundred microliters (380 μg; 40 μM) of a PEG-HoK solution (2 mg/mL) was added to 82 μg GEM-91 (10 μM) in 1 mL 10 mM Hepes buffer, pH 7.4. The solution was vortexed and kept for 30 min at 20°C. The size distribution of complexes was measured before and after addition of 30 μL of a 5 M NaCl solution. QELS measurements were performed as described in Fig. 4. PEG-HoK was prepared as followed: 2 mg HoK (0.39 μmol) dissolved in 40 μL DMSO in the presence of 7.8 μmol DIEA was reacted with 1.9 mg PEG-Osu (Mw 5000, Sigma) (PEG/HoK molar ratio of 1) in 10 μL DMSO. After 2 h at 20°C, 1 mL of distilled water was added.

and between ODN and pHPMA-b-pTMAEM block copolymers.[43] On the basis of Vinogradov et al.'s study,[42] a PEG-HoK/PS-ODN particle with an oligomer/ODN molar ratio of 3 could contain 12 PS-ODN and 37 PEG-HoK molecules. ODN/HoK complexes are cationic particles with a ζ potential of +17 mV, which drops to +2.5 mV upon PEGylation, similar to pDNA/HpK complexes.

ODN Transfer

HoK favors the cytosolic delivery of ODN from endosomes and increases their nuclear accumulation. Human nonsmall cell lung carcinoma cells (A549 cells) incubated with

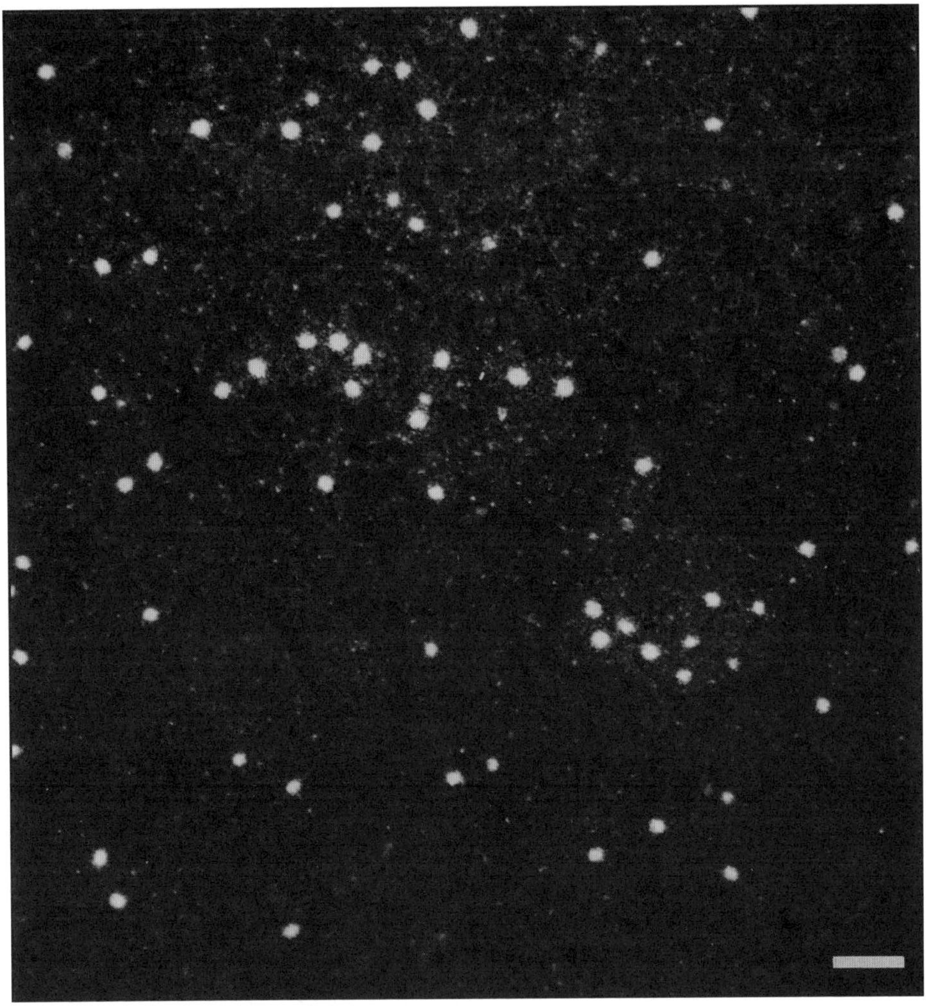

Fig. 3.11. Transmission Electron Microscopy of PEG-HoK/ODN complexes maintained 2 weeks at 4°C. The complexes were prepared as described in Fig. 3.10. The sample was processed and observed as described in the legend of Fig. 3.5. Scale bar = 100 nm.

Flu-PS-ODN in the absence of HoK exhibit faint vesicular staining (Fig. 3.12a), while their fluorescence intensity staining is stronger when ODN are complexed with HoK because of an enhancement of the ODN uptake (Fig. 3.12b). The cytosol and the nucleus are also strongly labeled. In contrast, the cell-associated fluorescence is low in the presence of AcHoK, which does not interact with ODN (Fig. 3.12c), as well as in the presence of histidine-free oligolysine (Fig. 3.12d). Double staining experiments using rhodamine-labeled HoK and fluorescein-labeled PS-ODN indicate that PS-ODN leave endosomes and reach the cell nucleus while HoK remains inside endosomes, probably entrapped in the membrane.[44] The effect of HoK is also observed for PO-ODN and the efficiency varies with the cell type.

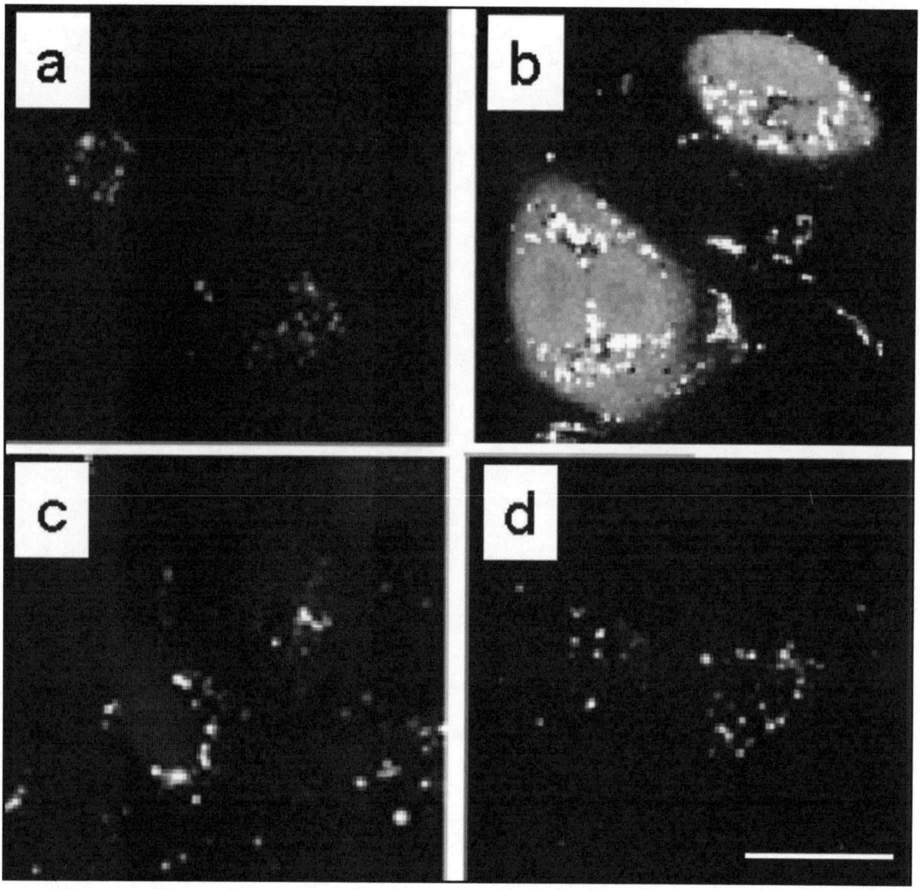

Fig. 3.12. HoK induces cytosolic and nuclear delivery of PS-ODN. A549 cells were incubated for 4 h at 37°C with *Flu*-PS-ODN (0.125 μM) either (a) in the absence of HoK, (b) in the presence of 20 μM HoK, (c) AcHoK, or (d) histidine-free oligolysine. Cells were fixed with 2% *p*-formaldehyde and visualized with a confocal microscope. Images were acquired by using identical settings. Scale bar = 25 μm. From Ref. (44).

HoK Increases the Biological Activity of Antisense PS-ODN

As expected, the increase in PS-ODN delivery in the cytosol with HoK correlates with an increase in their biological activity. In the presence of 20 μM HoK, the ICAM-1 expression on A549 cells induced by TNF-α is strongly inhibited (75%) by 1 μM ISIS 1939, an antisense PS-ODN complementary to the 3′ noncoding region[45] (Fig. 3.13). The inhibition is only 20% in the absence of HoK, and HoK by itself does not induce any inhibition of ICAM-1 expression. The inhibition is sequence-specific because GEM-91, a PS-ODN complementary to the AUG initiation site of $gag_{HIV\text{-}1}$ gene, induces only a limited inhibition (16%) in the presence of HoK (Fig. 3.13). An IC_{50} of 0.25–0.5 μM was found for ISIS 1939 in the presence of HoK, which is similar to that obtained with cationic lipids ($IC_{50} \sim 0.5$ μM).[45] or complexes with fucosylated polylysine in the presence of chloroquine ($IC_{50} \sim 0.5$ μM).[21]

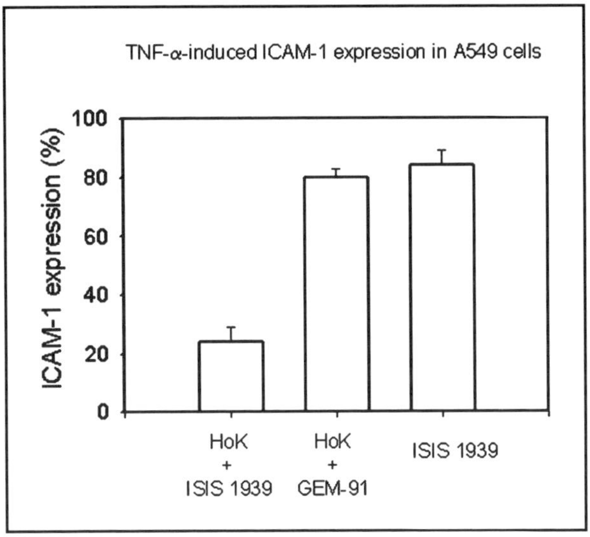

Fig. 3.13. Inhibition of TNF-α induced ICAM-1 expression by ISIS 1939. A549 cells were incubated at 37°C for 4 h in 100 μL DMEM serum-free medium containing 1 μM ISIS 1939 ODN either without or with 20 μM HoK. Cells were also incubated with 1 μM GEM91 ODN with HoK. One volume of fresh medium containing 10 ng/mL TNF-α was added, and cells were further incubated for 18 h. ICAM-1 expression was quantified by ELISA using anti-ICAM-1 antibodies. Data are means of triplicate experiments ± standard deviation of the percentage of control ICAM-1 expression induced by TNF-α. From Ref. (44).

Imidazole Protonation and Membrane Permeabilization

Polyhistidylated molecules, such as those described in this chapter, can induce membrane permeabilization upon the protonation of their imidazole rings in acidic medium. The permeabilization of the plasma membrane of nonadherent cells can be easily measured by flow cytometry using ethidium bromide (BET).[46] This nonfluorescent and poorly membrane-permeant molecule enters the cells upon membrane permeabilization and becomes strongly fluorescent upon binding to DNA. In the presence of 36 μM H5WYG or 20 μM HoK, 97% of the cells are labeled at pH 6.2 within 15 min, 50% at pH 6.8, and less than 5% at pH 7.1 (Fig. 3.14). H5WYG and HoK have a better permeabilizing capacity in slightly acidic medium than does the anionic peptide E5CA. Fifty percent of the cells are permeabilized at pH 6.0 with E5CA and none at pH 6.8. In contrast to E5CA which exhibits an α-helix structure in an acidic medium, H5WYG does not exhibit an ordered structure in acidic medium, suggesting a different membrane permeabilization mechanism.

H5WYG, HoK, and HpK containing several imidazoles are supposed to favor ODN and pDNA delivery in the cytosol after their protonation in endosomes. We have shown that the protonation of 50% of the imidazoles of HpK occurs at pH 6.9 versus pH 6.0 for histidine.[36] This pH elevation is due to the presence of the α-amino group of His. The imidazole protonation is also elevated when HpK is complexed with pDNA. Therefore, imidazole protonation can probably occur in endosomes at pH 6.0 and can mediate their destabilization to deliver pDNA and ODN in the cytosol. This is supported by the fact that the lack of endosome acidification by bafilomycin A_1, an inhibitor of the ATPase

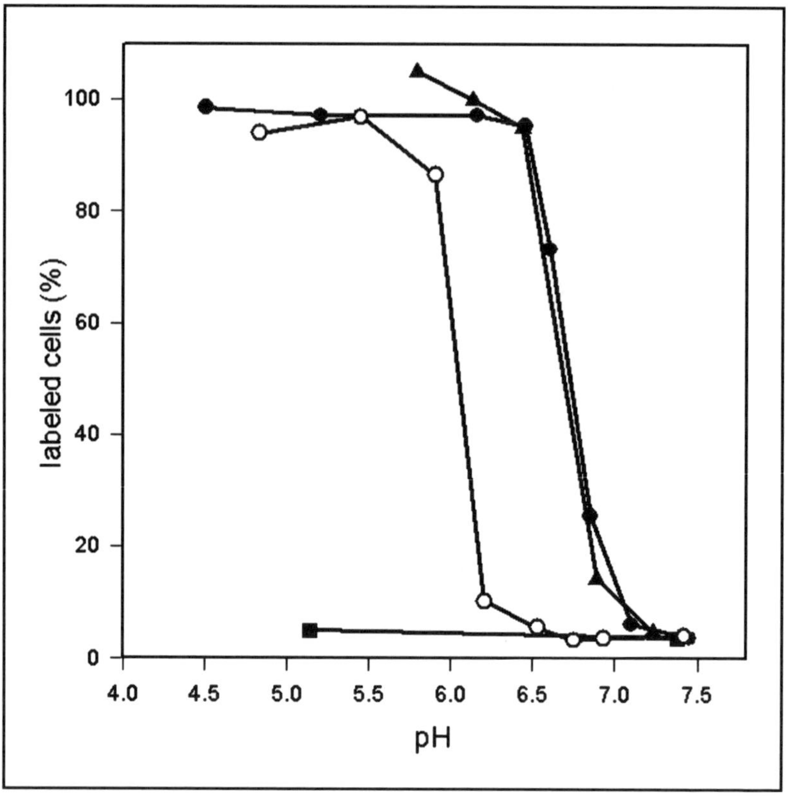

Fig. 3.14. pH-dependence of H5WYG- and HoK-induced membrane permeabilization. Nonadherent cells (5×10^5 cells) in 1 mL of sheath fluid (134 mM NaCl, 3.7 mM KCl, 15.2 mM NaF, 1.9 mM KH_2PO_4, 16.5 mM Na_2HPO_4, 0.2% 2-phenoxyethanol, pH 7.4) were incubated at 20°C with 12.5 μM BET in the absence (■) or in the presence of either (●) 36 μM H5WYG, (○) 36 μM E5CA, or (▲) 20 μM HoK. The pH of the cell suspension was lowered by addition of aliquots of a 0.3 N HCl solution. After 15 min, the red cell fluorescence intensity was measured by flow cytometry (λ exc. = 488 nm; λem = 585 ± 44 nm) and the percentage of red labeled cells was determined.

proton pump of endosomes[47] (Fig. 3.1b), drastically reduces the effect of H5WYG, HoK, and HpK. Polyfection with Lac-pLK/pDNA complexes in the presence of H5WYG as well as with HpK/pDNA complexes is strongly inhibited (82 and 84%, respectively) in the presence of bafilomycin A_1 (Table 3.3). The inhibition of the luciferase activity into pRETLuc cells by GEM-91, a PS-ODN complementary to the AUG initiation site of gag_{HIV-1} gene, with HoK is also decreased in the presence of bafilomycin A_1 (Table 3.3). These Rb-1 cells express stably the firefly luciferase reporter gene with the initiator AUG codon and several downstream codons of gag_{HIV-1} gene. In addition, confocal microscopy images confirm that bafilomycin A_1 treatment reduces drastically the amount of *Flu*-ODN in the cytosol and in the cell nucleus (Fig. 3.15).

Conclusion and Future Developments

In this chapter, we have described the effect of an histidine-rich peptide (H5WYG), histidylated oligolysine (HoK), and polylysine (HpK) on the cytosolic delivery and the

Table 3.3. Influence of endosome acidification on the effect of peptide, oligolysine, and polylysine containing histidine

	RLU × 10^{-3}		
	(a) Lac-pLK/pDNA (+H5WYG)	(b) HpK/pDNA	(c) HoK/ODN
−Bafilomycin A_1	46200 ± 9000	9000 ± 270	125 ± 12.5
+Bafilomycin A_1	380 ± 80	14 ± 4	217 ± 10
Inhibition (%)	82	84	42

Note. B16 cells were transfected with Lac-pLK/pCMVLuc complexes in the presence of 19 μM H5WYG peptide. (b) Rb-1 cells were transfected with HpK/pCMVLuc complexes. The transfection was conducted either in the absence or in the presence of 200 nM bafilomycin A_1. After 4h at 37°C, the medium was changed and cells were cultured for 48 h at 37°C. (c) pRET-Luc cells were incubated with 0.5 μM GEM-91 and 20 μM HoK2, either in the absence or in the presence of 800 nM bafilomycin A1. Then, the medium was removed and cells were incubated overnight in fresh complete medium containing 200 nM bafilomycin A1. The relative light units (RLU) represent the luciferase activity per cell. From Ref. (34, 41, 44).

biological activity of nucleic acids. HoK forms small (35 nm) cationic spherical particles with ODN, and HpK forms rod and toroid cationic particles (100 nm) with pDNA. These particles aggregate at physiological salt concentration, but interestingly for future developments, the stability, solubility, neutrality, and preservation of polyplexes are greatly improved by PEGylation. PEGylation, which reduces the nonspecific binding on the

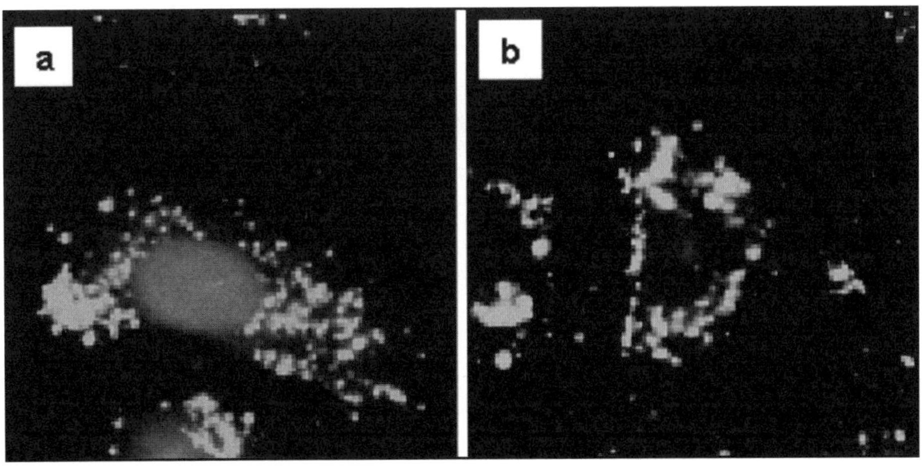

Fig. 3.15. Cytosol delivery of ODN requires endosome acidification. A549 cells were incubated for 4 h at 37°C with 0.1 μM *Flu*-PS-ODN and 20 μM HoK either (a) in the absence or (b) in the presence of 800 nM bafilomycin A1. Then, the medium was removed, and cells were further incubated for 30 min at 37°C in fresh complete medium containing 200 nM bafilomycin A1. Cells were fixed with 2% *p*-formaldehyde and visualized with a confocal microscope.

cell surface, opens the way for targeting of polyplexes upon linkage of neutral selective ligands such as peptides or sugars. On the basis of our findings, other histidine containing polycationic molecules such as *N*-Ac-poly(L-histidine)-graft-poly(L-lysine) comb shaped polymer[48] and a low molecular weight disulfide cross-linking peptide, Cys-His-$(Lys)_6$-His-Cys,[49] have been recently described for gene delivery.

While the delivery of ODN in the cytosol and then in the nucleus is clearly evidenced by confocal microscopy, more sensitive methods should be used for evaluating pDNA delivery. If it is clearly evidenced that in the presence of HoK, ODN alone are delivered in the cytosol, it is not known if either pDNA or polyplexes are delivered in the cytosol and imported in the nucleus. Recent data obtained with PEI-polyplexes and EGF-polylysine polyplexes seem to indicate that both pDNA and the cationic polymer can reach the nucleus.[50,51] The improvement of the efficiency of HpK requires the knowledge of the intracellular trafficking of polyplexes and pDNA including their transmembrane passage, their fate in the cytosol as well as the entry of pDNA in the nucleus. While ODN in the cytosol spontaneously diffuse into the nucleus, the foreign gene must be transported to the nucleus, released from the vector (if it is not yet accomplished in the cytosol) and must be expressed by the nuclear machinery to achieve its therapeutic effect. Despite several attempts performed with pDNA conjugated with peptide nuclear localization signal (NLS) molecule, such as that of the nucleocapside of the SV40 virus, the transfection efficiency has been only weakly improved.[52–54] This might suggest that other conjugation strategies or other types of NLS molecules should be explored, but also that other limiting steps have to be fully considered and optimized. Among them, the delivery of pDNA or polyplexes in the cytosol has to be unraveled. To date, little is known about the relationship between the number of pDNA copies available in the cytosol and the transfection efficiency. An interesting study has been reported, which indicates that the transfection is fully effective when 100 pDNA copies are injected directly in the nucleus whereas the injection of 10^5 pDNA copies in the cytoplasm leads to only the transfection of 25% of microinjected cells.[55] This low yield is not surprising; naked DNA does not diffuse in the cytosol and is digested by cytosolic nucleases.[33,56] The score increases up to 40% when 10^4 pDNA copies are injected in the cytoplasm upon condensation with PEI, which protects pDNA from nucleases. An another interesting study has also shown that 100% of the cells would be transfected if more than 10^6 pDNA copies complexed with a cationic lipid are internalized per cell.[57] These data indicate that the transfection efficiency is dependent not only on the amount but also on the fate of the plasmid in the cytosol. However, little is known about the effectiveness of plasmid delivery in the cytosol upon endocytosis of pDNA/HpK complexes. To achieve this, we have attempted to use cells stably transfected with the T7 RNA polymerase. Theoretically, the expression of a plasmid encoding reporter gene under the control of the T7 promoter, which does not require the nuclear machinery, will quantify the efficiency of the gene delivery in the cytosol. Unfortunately, this approach did not succeed because of the weakness of the T7 promoter and of the high stability of pDNA/HpK complexes that does not allow the transcription of pDNA. Methods which are able to measure the number of pDNA copies in the cytosol will be usefull. Real-time quantitative PCR, which is surely the best method, is actually performed to evaluate the number of pDNA copies in the cytosolic, nuclear, and vesicular extracts of transfected cells. These future investigations will clarify some early limiting steps of this attractive nonviral DNA delivery system.

Acknowledgments

This work was partly supported by grants from Agence Nationale de Recherche sur le Sida (ANRS), Association de Recherche sur le Cancer (ARC), EU (Bio4-CT97-2216), and Association Française de lutte contre la mucoviscidose (AFLM). We thank Brigitte

Guérin and Mahajoub Bello Roufai for their contributions. CBM is affiliated with the University of Orléans and with INSERM.

References

1. Rolland AP. Crit Rev Ther Drug Carrier Syst 1998; 15:143-198.
2. Zauner W, Ogris M, Wagner E. Adv Drug Delivery Rev 1998; 30:97-113.
3. Garnett MC. Crit Rev Ther Drug Carrier Syst 1999; 16:147-207.
4. Fajac I, Allo J-C, Souil E et al. J Gene Med 2000; 2:368-378.
5. Luthman H, Magnusson G. Nucleic Acids Res 1983; 11:1295-1308.
6. Zenke M, Steinlein P, Wagner E et al. Proc Natl Acad Sci USA 1990; 87:3655-3659.
7. Hasan MT, Subbaroyan R, Chang TY. Somat Cell Mol Genet 1991; 17:513-517.
8. Midoux P, Mendes C, Legrand A et al. Nucleic Acids Res 1993; 21:871-878.
9. Erbacher P, Roche AC, Monsigny M et al. Exp Cell Res 1996; 225:186-194.
10. Wagner E, Plank C, Zatloukal K et al. Proc Natl Acad Sci USA 1992; 89:7934-7938.
11. Haensler J, Szoka FC. Bioconjugate Chem 1993; 4:372-379.
12. Plank C, Oberhauser B, Mechtler K et al. J Biol Chem 1994; 269:12918-12924.
13. Zauner W, Blaas D, Kuechler E et al. J Virol 1995; 69:1085-1092.
14. Gottschalk S, Sparrow JT, Hauer J et al. Gene Ther 1996; 3:448-457.
15. Wyman TB, Nicol F, Zelphati O et al. Biochemistry 1997; 36:3008-3017.
16. Kichler A, Freulon I, Boutin V et al. J Gene Med 1999; 1:134-143.
17. Hélène C, Toulme JJ. Biochim Biophys Acta 1990; 1049:99-125.
18. Wu GH, Wu CH. J Biol Chem 1992; 267:12436-12439.
19. Citro G, Perrotti D, Cucco C et al. Proc Natl Acad Sci USA 1992; 89:7031-7035.
20. Citro G, Szczylik C, Ginobbi P et al. Br J Cancer 1994; 69:463-467.
21. Stewart AJ, Pichon C, Meunier L et al. Mol Pharmacol 1996; 50:1487-1494.
22. Mahato RI, Takemura S, Akamatsu K et al. Biochem Pharmacol 1997; 53:887-995.
23. Dheur S, Dias N, van Aerschot A et al. Antisense Nucleic Acid Drug Dev 1999; 9:515-525.
24. Derossi D, Joliot AH, Chassaing G et al. J Biol Chem 1994; 269:10444-10450.
25. Bongartz JP, Aubertin AM, Milhaud PG et al. Nucleic Acids Res 1994; 22:4681-4688.
26. Bergan R, Hakim F, Schwartz GN et al. Blood 1996; 88:731-741.
27. Gao WY, Storm C, Egan W et al. Mol Pharmacol 1993; 43:45-50.
28. Temsamani J, Kubert M, Tang J et al. Antisense Res Dev 1994; 4:35-42.
29. Chin DJ, Green GA, Zon G et al. New Biol 1990; 2:1091-1100.
30. Leonetti JP, Mechti N, Degols G et al. Proc Natl Acad Sci USA 1991; 88:2702-2706.
31. Fisher TL, Terhorst T, Cao X et al. Nucleic Acids Res 1993; 21:3857-3865.
32. Spiller DG, Tidd DM. Antisense Res Dev 1995; 5:13-21.
33. Lukacs GL, Haggie P, Seksek O et al. J Biol Chem 2000; 275:1625-1629.
34. Midoux P, Kichler A, Boutin V et al. Bioconjugate Chem 1998; 9:260-267.
35. Pichon C, Freulon I, Midoux P et al. Antisense Nucleic Acid Drug Dev 1997; 7:335-343.
36. Bello-Roufai M, Midoux P. Bioconjugate Chem 2001; 12:92-99.
37. Ogris M, Brunner S, Schuller S et al. Gene Ther 1999; 6:595-605.
38. Erbacher P, Bettinger T, Belguise-Valladier P et al. J Gene Med 1999; 1:210-222.
39. Delain E, Le Cam E. In Visualization of Nucleic Acids, G Morel (ed.), CRC Press, Boca Raton, FL, 1995, pp. 35-56.
40. Erbacher P, Roche AC, Monsigny M et al. Biochim Biophys Acta 1997; 1324:24-36.
41. Midoux P, Monsigny M. Bioconjugate Chem 1999; 10:406-411.
42. Vinogradov SV, Bronich TK, Kabanov AV. Bioconjugate Chem 1998; 9:805-812.
43. Read ML, Dash PR, Clark A et al. Eur J Pharm Sci 2000; 10:169-177.
44. Pichon C, Bello-Roufai M, Monsigny M et al. Nucleic Acids Res 2000; 28:504-512.
45. Chiang MY, Chan H, Zounes MA et al. J Biol Chem 1991; 266:18162-18171.
46. Midoux P, Mayer R, Monsigny M. Biochim Biophys Acta 1995; 1239:249-256.
47. Bowman EJ, Siebers A, Altendorf K. Proc Natl Acad Sci USA 1988; 85:7972-7976.
48. Benns JM, Choi JS, Mahato RI e al. Bioconjugate Chem 2000; 11:637-645.
49. McKenzie DL, Smiley E, Kwok KY et al. Bioconjugate Chem 2000; 11:901-909.
50. Godbey WT, Wu KK, Mikos AG. Proc Natl Acad Sci USA 1999; 96:5177-5181.

51. Schaffer DV, Fidelman NA, Dan N et al. Biotechnol Bioeng 2000; 67:598-606.
52. Branden LJ, Mohamed AJ, Smith CI. Nat Biotechnol 1999; 17:784-787.
53. Zanta MA, Belguise-Valladier P, Behr JP. Proc Natl Acad Sci USA 1999; 96:91-96.
54. Neves C, Byk G, Scherman D et al. FEBS Lett 1999; 453:41-45.
55. Pollard H, Remy JS, Loussouarn G et al. J Biol Chem 1998; 273:7507-7511.
56. Lechardeur D, Sohn KJ, Haardt M et al. Gene Ther 1999; 6:482-497.
57. Tseng WC, Haselton FR, Giorgio TD. J Biol Chem 1997; 272:25641-25647.

CHAPTER 4

Intracellular Targets for DNA Delivery: Nuclei and Mitochondria

Vladimir P. Torchilin, Ban-An Khaw, and Volkmar Weissig

Abstract

All discussions on the intracellular delivery of DNA are based on a seemingly evident assumption that the key task is to bring the intact DNA into the cell cytoplasmic compartment, and then the DNA will find its way to a right place. The nuclear genome is usually considered to be this "right place." However, until recently, in numerous experiments on the intracellular DNA delivery, it has been almost completely neglected that cells contain another genome, the mitochondrial one. And, in many cases, this genome should become a therapeutic target. Being delivered inside the cell, DNA actually has two ways to go—to nuclei and to mitchondria, and the proper choice between these two ways may be decisive for the success of gene therapy. Certainly, nuclear DNA delivery is far more advanced than mitochondrial delivery one. In addition, free DNA from the cytoplasm has a strong tendency to spontaneously associate with the nuclear genome. Mitochondria as a target for DNA have much less accessibility, still remaining an important site to reach. Whereas the nuclear delivery of DNA is under active investigation and just awaits better protocols to be elaborated, practically applicable mitochondrial DNA delivery is at its early stage and must be developed almost from scratch. In our studies on intracellular DNA delivery, we have attempted to develop new protocols for targeting DNA to nuclei and to mitochondria. In this chapter we provide a brief description of our recent experiments in both of these important areas.

Abbreviations: CSIL, cytoskeleton-specific immunoliposomes; DQA, dequalinium; IL, immunoliposomes; PL, plain liposomes; IgL, liposomes with nonspecific immunoglobulin; OG, octylglucoside; 3HT, 3H-thymidine; RLU, relative light unit; PBS, phosphate-buffered saline; ATP, adenosine triphosphate; mtDNA, mitochondrial DNA; CPM, cytoplasmic membrane-like liposomes; IMM, inner mitochondrial membrane-like liposomes; OMM, outer mitochondrial membrane-like liposomes.

Synthetic DNA Delivery Systems, edited by Dan Luo and W. Mark Saltzman.

Intracytoplasmic Gene Delivery with Cytoskeleton-Specific Immunoliposomes

DNA Delivery and Cytoskeleton-Specific Immunoliposomes

Gene therapy is now considered by many as a major contender in the future treatment protocols for various diseases, including cancer[1] and AIDS.[2] Although some early initial trials were encouraging, the latest results have raised certain concerns.[3] Viral vectors for DNA delivery suffer from nonspecificity and inherent risks of virus-induced complications. Nonviral delivery systems, first of all, cationic lipids/liposomes,[4] also have certain drawbacks, such as the same nonspecificity and cytotoxic reactions,[5–7] though new cationic lipid derivatives with decreased toxicity are currently under development.[8,9] To minimize potential problems, in vitro transfection, with subsequent introduction of the transfected cells or organs such as vein grafts,[10] donor hearts,[11] pulmonary artery segments,[12] and lung transplants[13] into recipients, was suggested. Fibroblasts transfected in vitro with epidermal growth factor gene have already been successfully used for wound healing.[14] Still, the traditional routes of internalization of the DNA carriers by endocytosis or pinocytosis, with subsequent degradation of the delivered DNA by lysosomal nucleases, strongly limit the efficacy of transfection.[15] From this point of view, the development of a new method that can in vitro deliver genetic constructs directly into the cytoplasm of the target cells would be highly desirable.

Earlier, cytoskeletal antigen (myosin) specific immunoliposomes (CSIL) were shown to seal membrane lesions in hypoxic cardiocytes by anchoring CSIL to the exposed cytoskeletal antigen via the antimyosin antibody, which results in membrane repair and restoration of cell viability.[16] Since some of the adhered liposomes were shown to fuse with cells and deliver the intraliposomal contents directly into the cell cytoplasm, this approach may also be used for targeted DNA delivery into the cell cytoplasm bypassing the endocytotic route,[17,18] as shown in the principal scheme in Fig. 4.1.

Fusion of CSILs with Hypoxic Cells and Delivery of Liposomal Contents into Cell Cytoplasm

Antimyosin antibody, a marker of an intracellular cytoskeletal antigen, has been used in the past to demonstrate sacrolemmal lesions typical of necrotic myocyte damage.[19,20] Hypoxia-provoked membrane disruption cannot be reversed by simple restoration of blood flow to the ischemic zone. We hypothesized that if the membrane lesions could be sealed to prevent loss of intracellular molecules, then the treated cells should recover and remain viable. With this in mind, we proposed the use of antibody-targeted liposome-plugs,[16–18] wherein a cytoskeletal antigen exposed via membrane lesion can be used to anchor the immunoliposome to provide the initial protection (seal) and prevent the leakage of the intracellular contents.

The applicability of the "plug and seal" concept to prevent hypoxia-mediated cell death was demonstrated using myosin as the cytoskeletal target antigen and the corresponding antimyosin antibody as the anchoring device incorporated in the liposomes, and tested in a hypoxic model of injury of H9C2 rat embryonic cardiocytes.[16] H9C2 cardiocytes, subjected to temporary artificial hypoxia, were incubated with antimyosin immunoliposomes (IL), plain liposomes (PL), and liposomes with an attached nonspecific control IgG (IgL). To prepare CSIL, monoclonal antimyosin antibody was modified with the hydrophobic group of *N*-gluaryl phosphatidyl ethanolanime as reported earlier.[16]

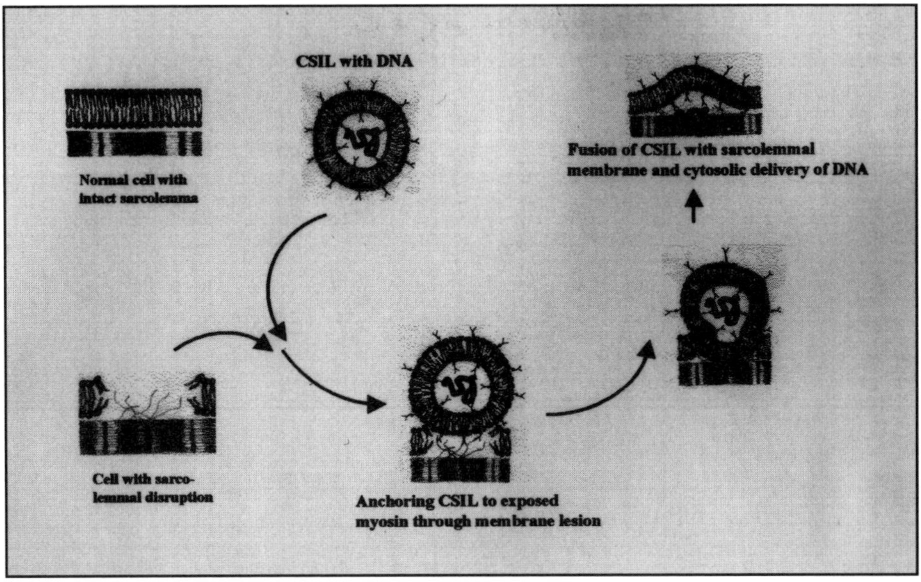

Fig. 4.1. Diagrammatic representation of the mechanism of CSIL-fection. The molecule in the liposome represents the DNA to be inserted into the cardiocyte that has undergone membrane damage. The location of the DNA molecule relative to the liposome is arbitrary. Adapted from Ref. (24).

Liposomes were prepared from phosphatidyl choline, cholesterol, and modified protein mixtures by the detergent (octylglucoside, OG) dialysis method.

The assessment of cell viability was performed after 24 h of hypoxia by the trypan-blue exclusion method or by cell incubation with [^{3}H]-thymidine (3HT). According to the trypan-blue dye exclusion data, almost all control hypoxic cardiocytes were nonviable. Plain liposomes and IgL provided some protection from hypoxic injury, probably, by nonspecifically sticking to cell surfaces and fortuitously "sealing" some of the cell membrane breaches. However, CSIL completely prevented cell death and provided cell viability similar to that of normoxic cells. 3HT uptake studies indicative of DNA replication (and cell viability) demonstrated a similar pattern. If the mean 3HT uptake in control normoxic cardiocytes was taken as 100%, the uptake in untreated hypoxic controls was only ca. 3% after 24 h of hypoxia. At the same time, the uptake in CSIL-treated hypoxic cells was almost 90% of normoxic cells and significantly greater than in PL-treated hypoxic controls (ca. 30%).

The set of experiments was performed to show that cell salvage is, at least in part, due to the fusion of the IL lipid bilayer with the cell membrane promoted by the anchoring of the CSIL to the underlying cytoskeletal myofilaments, rather than simple plugging of the sarcolemmal lesions with CSIL. With this in mind, we have attempted to learn more about the exact mechanism of CSIL interaction with hypoxic cardiocytes by performing ultrastructural studies of hypoxic cardiocytes treated with CSIL or PL. To demonstrate that CSIL deliver the intraliposomal contents into the cytoplasm of the cardiocytes, the method previously described by Torchilin et al.[21] was employed[16] using silver/silver-oxide-grains-containing liposomes.

We had initially shown that these silver grains entrapped into CSIL were delivered in significant quantities into the cytoplasm of hypoxic cardiocytes examined by transmission electron microscopy.[16] Electron microscopy of hypoxic cells incubated with silver-oxide-loaded CSIL demonstrated internalization of the silver grains and localization in the

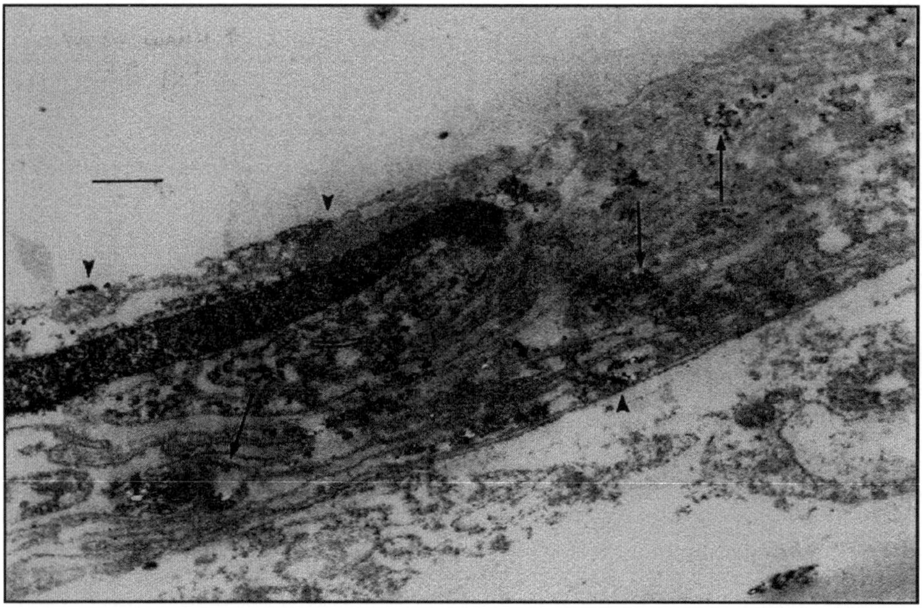

Fig. 4.2. Transmission electronmicrograph of a hypoxic cardiocyte treated with silver-grains-loaded CSIL after 24 h of hypoxia. The silver grains are observed in the intracytoplasmic space (thin arrows) or on the outer surface of the plasmic membrane (thick arrows). They are not enclosed in endosomal vesicles. Bar = 1 μ. Adapted from Ref. (16).

cytoplasmic compartment (see Fig. 4.2). Figure 4.2 shows the transmission electron micrograph of a hypoxic cardiocyte treated with antimyosin CSIL with entrapped silver grains. As can be seen, the silver grains in this electronmicrograph are located in the cytoplasm and do not appear to be enclosed in endosomes or lysosomes, denoting possible fusion of the CSIL with the cell membrane followed by the release of the intraliposomal contents into the cytoplasm.[16,21] The hypoxic cardiocytes treated with silver-grains-containing PL show no internalization of the silver grains. We have concluded from this that artificially imposed hypoxia may be used to facilitate intracellular delivery of various drugs including DNA constructs by CSIL. If the target cells for drug and/or gene delivery are under artificially imposed hypoxic stress, stress-induced small membrane lesions will allow intracellular trafficking of liposomes rendered specific for an intracellular antigen. Such liposomes may simultaneously plug and seal hypoxia-induced cell membrane lesions and provide intracellular drug (gene) release and delivery. For in vitro targeted gene transfection, transient, reversible hypoxic injury may be induced in the target cells or tissue grafts, and DNA-bearing CSIL deliver the constructs directly into the cytoplasm with restoration of the membrane integrity and cell viability. We named this method "CSIL-fection."

Intracytoplasmic Delivery of DNA by Cytoskeleton-Specific Immunoliposomes

The hypothesis on the possibility of using CSIL-fection for intracytoplasmic delivery of DNA was verified in tissue cultures utilizing H9C2 rat embryonic cardiocytes as target cells and CSIL as carriers for targeted gene transfection containing various reporter genes. To prepare DNA-loaded liposomes, corresponding plasmids pEScFv were added to OG-solubilized liposomal lipids before final detergent removal by dialysis (when necessary, together with modified antibody). The liposome-bound DNA was separated from free DNA by Ficoll gradient centrifugation. The presence of a substantial fraction of added DNA (up to 50% in some runs) in the liposome fraction was confirmed by agarose gel electrophoresis of original DNA, and liposome samples before and after Ficoll gradient separation. H9C2 rat embryonic cardiocytes prepared as described earlier[16] were grown in 25 mL flasks, in fresh medium containing DNA-loaded CSIL, IgL, or PL or in fresh medium with or without naked DNA, and subjected to flushing with sterile N_2 gas for 4 min. The flasks were closed tightly and hypoxia maintained for 24 h. Normoxic controls were included. The media were removed and cells washed with fresh medium. Fresh medium was added and incubation continued for 48 h. All studies were performed in sets of six and repeated three times.

The initial confirmation study was performed with a plasmid pEScFv 2G42D7 vector.[22] Plasmid pEScFv is an eukaryotic expression vector containing the antimyosin single chain Fv (ScFv) fragment linked with domain B of protein A of *Staphylococcus aureus*. Sp2/O myeloma cells transfected with pEScFv 2G42D7 secreted ScFv into the culture media. *Escherichia coli* strain XL1/Blue was transformed with pEScFv 2G42D7 plasmid, and plasmid DNA was obtained by alkaline lysis method[23] from the ampicillin-resistant line. H9C2 cells were subjected to hypoxic stress in the presence of CSIL–plasmid complex. ELISA assays with rat cardiac myosin were performed to assess the expression of 2G42D7 ScFv by CSIL-fected cells after their incubation for another 48 h under the normoxic conditions. The data obtained clearly demonstrate the efficacy of the CSIL-fection process: the transfection efficiency of hypoxic cells with CSIL was 3.5 times higher than for PL with hypoxic cells and 5.5 times higher than for PL with normoxic cells; see Table 4.1. (Better expression in PL-treated hypoxic cells than in PL-treated normoxic cells can be explained by increased plasmid delivery in hypoxic cells nonspecifically sealed with PL.)

The efficiency of transfection with luciferase pGL2 vector using CSIL, PL, and naked DNA in hypoxic or normoxic H9C2 cardiocytes[24] is shown in Fig. 4.3. With pGL2 vectors, the cells were collected, lysed, and the luciferase activity assessed by using the Luciferase Assay Reagents and relative light (intensity) unit (RLU) measured using a luminometer. Only hypoxic cells treated with pGL2–CSIL-fection showed the expression of luciferase activity (ca. 72,000 RLU). Normoxic H9C2 cells treated with CSIL-fection or PL, hypoxic H9C2 treated with PL, normoxic cells treated with naked pGL2 DNA or no DNA all showed similar background activity. CSIL-fection was also compared to cationic liposome-mediated transfection with the same quantity of DNA-loaded cationic

Table 4.1. Expression of ScFv in transfected cardiocytes

Cells and liposomes	**ScFv concentration by ELISA (in μg/mL antibody concentration)**
Normoxic cells + PL	ca. 0.09
Hypoxic cells + PL	ca. 0.21
Hypoxic cells + CSIL	ca. 0.48

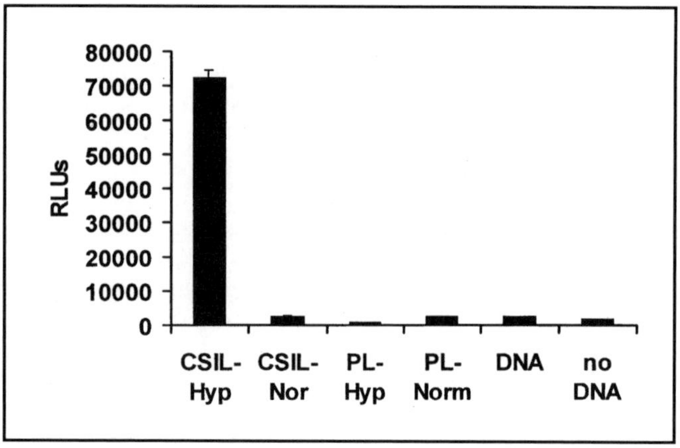

Fig. 4.3. Relative light units (RLUs) produced by firefly luciferase as evidence of the efficiency of transfection by CSIL or PL in normoxic and hypoxic H9C2 cardiocytes. The expression of the transfected genes was also compared to normoxic cells treated with naked DNA and no DNA. From Ref. (24).

liposomes (as lipid). The RLU generated in cells treated with cationic liposomes was ca. 380,000 compared with ca. 72,000 RLUs produced by the same number of cells treated with CSIL, or cationic liposomes were ca. 5 times more effective than CSIL-fection using pGL2 as reporter genes. However, cationic liposomes were shown to carry and deliver to cells much more DNA than we have found for CSIL.[25,16] Thus, simple calculations show that cationic liposomes deliver to cells ca. 10^6 times more DNA than does CSIL. It means, that on an equimolar basis of DNA delivery, CSIL-fection appears to be more efficient, i.e. much smaller number of CSILs and CSIL-associated DNA are required to produce 1 RLU by CSIL-fection.

We assumed that the higher net RLU generated by cationic liposome transfection than by CSIL-fection with pGL2 vector was due to a higher vector concentration in a fewer number of cardiocytes transfected with cationic liposomes than with CSIL rather than a greater number of cardiocytes successfully transfected. To demonstrate it and visualize the individual cells successfully transfected by either cationic liposomes or CSIL, studies with pSV-β-galactosidase vector were undertaken. The number of cells (in each microscopic field of view at ×100 magnification) CSIL-fected with pSV-β-galactosidase vector was compared to those with cationic liposomes, naked DNA, PL, IgL, and no DNA. The cells were fixed with 2% glutaraldehyde, washed with PBS, and color was developed with a commercial X-Gal substrate.[24] Figure 4.4a shows H9C2 cardiocytes successfully transfected with pSV-β-galactosidase-CSIL-fection. The blue color (more dark cells on b/w pictures) developed with the substrate X-Gal at pH 7.4 is specific for the expressed bacterial β-galactosidase enzyme, and vector expression can be seen in almost all cells in the culture. In case of cationic liposomes, even with the optimal amount of vector (as indicated by the package insert for preparing Transfast™), an average of only 2–3 cardiocytes in each microscopic field were observed to be transfected (Fig. 4.4b). At the same time, the intensity of expression of β-galactosidase in cells transfected with cationic liposomes was much greater than of those in cells transfected with CSIL. No expression of β-galactosidase activity was seen with IgL treatment as well as in cells treated with naked DNA, hypoxic or normoxic cells treated with PL, normoxic H9C2 cells treated with CSIL, or hypoxic cells treated with IgL. Quantitation of the in vitro transfected H9C2 cardiocytes showed that CSIL-fection was able to transfect about 40 times more cells than cationic liposomes per microscopic field of view.

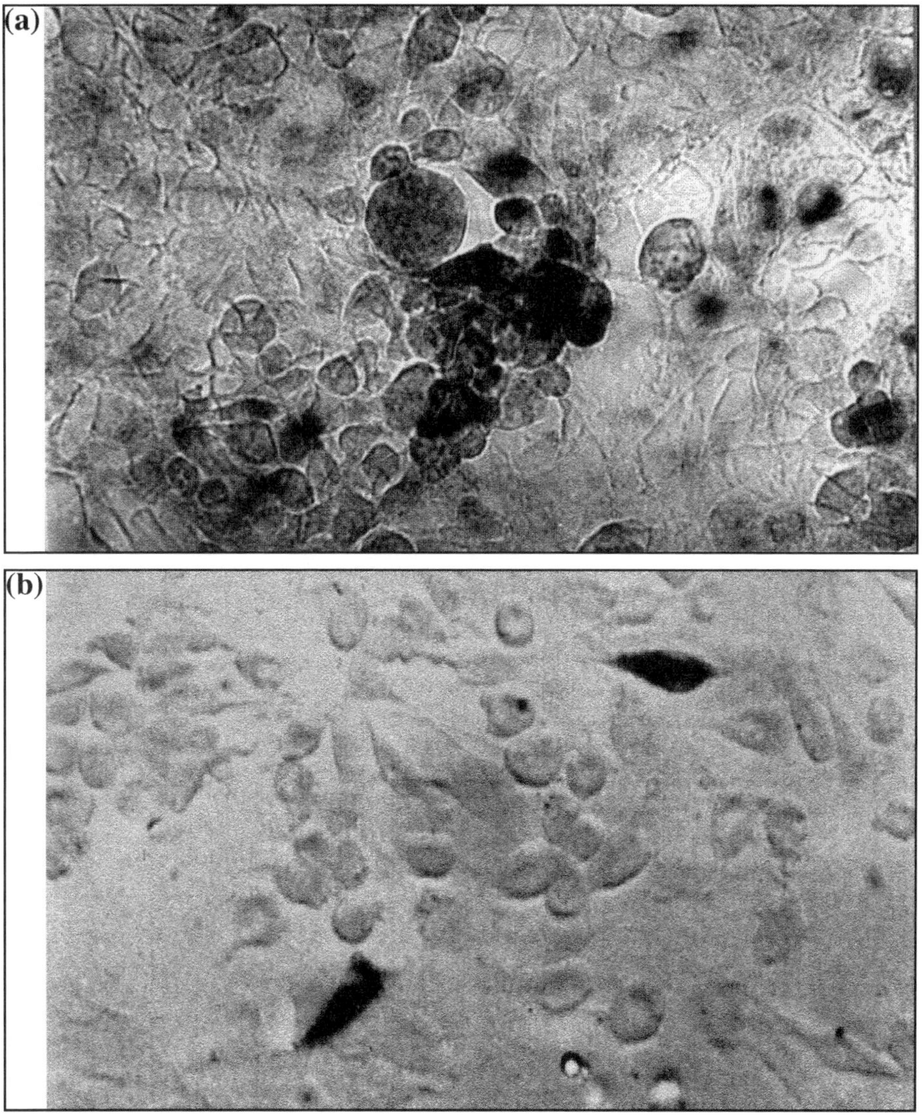

Fig. 4.4. (a) Photomicrograph of H9C2 cardiocytes CSIL-fected with pSV reporter genes. Gene expression (originally, blue color from β-galactosidase substrate; dark color on b/w image) in virtually all cells; (b) Photomicrograph of H9C2 cardiocytes transfected with cationic liposome(Lipofectin)-pSV-β-galactosidase reporter genes showing only few cells with intensive dark color (blue dye on the original image). Magnification ×400. Modified from Ref. (24).

Future Opportunities for the Application of CSIL-Fection

The CSIL-fection experiments with the three reporter genes, described above, clearly indicate that the novel method of gene transfection by direct intracytoplasmic delivery of the vectors utilizing CSIL and transitional and reversible artificial hypoxia is highly effective and quite comparable with existing nonviral methods, such as transfection with the use of cationic liposomes. Our data with pGL2 and pSV-β-galactosidase vectors showed that the net number of cells transfected by CSIL-fection is higher than that with cationic liposomes under comparable conditions, though the number of foreign DNA delivered into the cell is higher in the case of cationic liposomes (if cationic liposome-mediated transfection was achieved for a given cell at all). A possible explanation for transfection of the higher number of cells with CSIL-fection than with cationic liposomes may be that when a large number of cationic liposomes interact with one cell, there is a potential for interference of the excessive amounts of cationic lipids with the normal cell membrane integrity (a kind of "cooperative" cell permeabilization by cationic liposomes and subsequent delivery of higher quantities of DNA). On the other hand, the large number of cationic liposomes interacting (fusing) with one cell could also cause cell attrition or interfere with normal cellular functions. This is in contrast to cell membrane sealing with a small number of CSIL that are made of neutral lipids and therefore are not expected nor observed to cause adverse effects on normal cell function. From the quantitation of the number of fluorescent-CSIL, just ca. 100 intact CSIL were observed to be attached to a single cardiocyte at 24 h of hypoxic culture. At the same time it was found that up to 40,000 cationic liposomes can attach to a single cell in the process of cationic liposome-mediated transfection.[25] It was also observed that the cell membranes of these cardiocytes have acquired and incorporated some fluorescent lipids that were originally associated with the liposomes. Incorporation of the liposome lipid bilayer into the cell membrane is consistent with fusion of CSIL with the cell membrane.

The structure of association of DNA to liposome may also influence the efficiency of transfection. DNA may be entrapped into the liposome or form aggregates. The sizes of pSV-β-galactosidase and pGL2 are about 6000 bp. Since each bp is 3.35-Å,[26] in the linear form, the circular vectors have an approximate diameter of 700 nm and, therefore, cannot be located in the intraliposomal compartment (in our experiments, the size of liposomes was less than 300 nm). However, the majority of purified vectors that are not nicked with nucleases, exist as supercoiled plasmids, the diameter of which is only about 1/25 of the diameter of the corresponding open circular plasmids.[26] Assuming these calculations, the diameter of the supercoiled pSV-β-galactosidase and pGL2 plasmids should be only about 28 nm, and plasmids may be entrapped into the intraliposomal space. Even if CSIL and DNA form certain aggregates, they may still be able to attach at the cell membrane lesions, seal them, and simultaneously facilitate intracytoplasmic DNA delivery. The silver grains do not appear to be localized in structures consistent with endosomes or lysosomes, but appear to have a cytoplasmic localization.

The initial proof of feasibility of this gene transfection approach is shown with cardiocytes, however, by using a specific combination of another intracellular cytoskeletal antigen and the corresponding antibody. This method of transfection should also be applicable to other cell types or cellular gene therapy of organ systems. Furthermore, one can speculate that the CSIL-fection should be amenable to gene vaccination since it should be possible to induce a very small, localized area of skeletal muscle injury into which CSIL with the gene for vaccine are injected. The skeletal muscle cells with induced injury should undergo cell membrane lesion sealing with CSIL and thereby achieve intracytoplasmic delivery of the gene for vaccine. Since no viruses are involved in this transfection, the potentials for initiation of an immune response that can eliminate the transfected cells are not expected. Our in vitro method of intracytoplasmic gene delivery, followed by in vivo administration of the genetically transfected cells, should also provide a highly efficient approach combining the best of the in vitro to in vivo gene therapy.

Furthermore, in myocardial ischemia, toxic or inflammatory insult to the myocardium results in the loss of sarcolemmal integrity leading to the exposure of intracellular myosin to the extracellular milieu.[27,28] Earlier, targeting of the infarcted myocardium has been achieved with radiolabeled antimyosin immunoliposomes in rabbits and dogs.[29,30] Therefore, one may hypothesize that if CSIL were loaded with an angiogenic vector such as gene for VEGF, and are delivered to the compromised cardiocytes during the very acute phase of cell membrane lesion evolution, concurrent with thrombolytic therapy, CSIL should be able to seal the lesions and simultaneously induce neoangiogenesis.

In conclusion, a novel and highly efficient method of in vitro gene transfection employing direct intracytoplasmic gene delivery with neutral CSIL under reversible hypoxic conditions was developed in an embryonic cardiocyte model. Enhanced efficiency of transfection by this novel method could have practical therapeutic applications in genetic modification of cells ex vivo that could then be reimplanted into patients for gene therapy.

New Approach to DNA Delivery in Mitochondria with Mytochondria-Specific Carrier

Mitochondrial Diseases and Possibilities of Mitochondrial Gene Delivery

Other important targets for intracellular DNA delivery are mitochondria. The major function of mitochondria that are present in all eukaryotic cells, is the production of ATP by oxidative phosphorylation. Mitochondria are also involved in other metabolic pathways, such as the citrate cycle, the oxidation of fatty acids, and the synthesis of steroid hormones and gluconeogenesis. Each mitochondrion is composed of two membranes, which together create two separate compartments, the internal matrix space and the narrow intermembrane space. The outer membrane is permeable to all molecules smaller than 5 kDa, making the intermembrane space chemically equivalent to the cytosol with respect to small molecules. The inner membrane is highly specialized and characterized by an unusually high content of membrane proteins, and is highly impermeable even to very small molecules. This impermeability is prerequisite for the establishment of an electrochemical gradient, i.e. an imbalance in the distribution of protons, which drives the synthesis of ATP.

With the development of understanding molecular mechanisms of various pathologies, it has become evident that mitochondrial dysfunction contributes to a variety of human disorders, ranging from neurodegenerative diseases, obesity, and diabetes to ischemia-reperfusion injury and cancer. The number of diseases found to be associated with defects of the mitochondrial genome has grown significantly over the last decade.[31,32] Among the first mitochondrial diseases to be understood at the molecular level was a group of neuromuscular diseases resulting from mitochondrial DNA (mtDNA) deletions.[33] Large deletions as well as point mutations of mtDNA were found to lead to the whole variety of mitochondrial myopathies.[34] Despite major advances in understanding mtDNA defects at the genetic and biochemical level, there is no satisfactory treatment for the vast majority of patients available. However, mitochondrial gene therapy is currently only at the very early stage of its development.[35] In particular, any possibility for gene therapy requires the use of a yet unavailable mitochondrial transfection vector.[36]

One can think of two different approaches to mitochondrial gene therapy,[37] see Fig. 4.5. The first involves expressing a wild-type copy of the defective gene in the nucleus, with cytoplasmic synthesis and subsequent targeting of the gene product to the mitochondria. However, besides the different codon usage in mitochondria, there exists a set of limitations in adapting this nuclear–cytosolic approach for mammalian cells. First, the majority of mtDNA defects involve tRNAs and to date, no mechanism for the mitochondrial

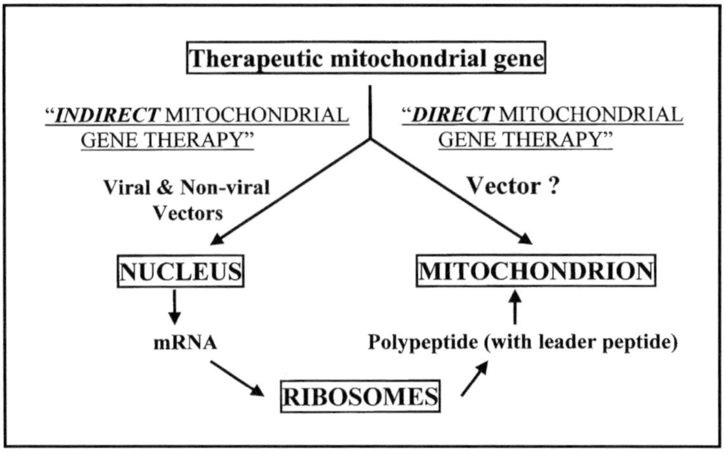

Fig. 4.5. Schematic illustration of the two strategies for mitochondrial gene delivery. The indirect way involves the nuclear–cytosolic protein synthesis pathway followed by mitochondrial leader peptide mediated uptake into the organelle. The direct way is based on the delivery of DNA directly into the mitochondria followed by transcription and translation at mitochondrial ribosomes. From Ref. (52).

uptake of cytosolic tRNAs in mammalian cells has been reported. Then, the 13 proteins encoded for by mtDNA are highly hydrophobic. It has therefore been suggested that these proteins need to be synthesized at the location of their function. And lastly, it has been hypothesized that some of the proteins encoded by the mitochondrion may be potentially toxic if synthesized in the cytosol.[38] Considering these problems associated with the nuclear–cytosolic approach, the direct transfection of mitochondria appears as a better alternative. It is timely to mention here that human mitochondrial DNA is a circular molecule 16,569 bp in size. Depending on the energy needs of the particular tissue, human cells contain between a few dozen to several thousand molecules of mtDNA.[39] A striking feature of human mtDNA is the lack of introns; the genes are generally joined end-to-end or separated by only a few nucleotides; noncoding stretches are missing.[39]

Several different approaches for the direct transfection of mitochondria have been described (reviewed in Ref. (40)). Although foreign DNA has been successfully introduced into mitochondria of yeast by biolistic bombardment of cells with DNA-coated tungsten particles,[41] this technique has not worked so far in mammalian cells. Plasmid DNA was also introduced into the matrix compartment of isolated mitochondria by electroporation.[42] However, if successful expression of the foreign gene can be shown, the question of whether mitochondria containing exogenous DNA can be introduced back into cells in vitro and in vivo to correct mitochondriopathies still has to be answered.[40] The correction of a mitochondrial deficiency in a cell model by transfer of "healthy" mitochondria through the cytoplast fusion in vitro was demonstrated in.[39] Few studies demonstrate the feasibility of delivering exogenous DNA–peptide conjugates into the mitochondrial matrix;[36,43] however, they do not address how these conjugates will pass the cytosolic membrane and reach mitochondrial proximity. Both studies were carried out only with isolated and energized mitochondria. So the question emerges whether classical transfection vectors have the potential to transport a mitochondrial leader peptide–DNA conjugate to the site of mitochondria in living cells.

All viral vectors transport DNA to the nucleus by virtue of their nature and nonviral DNA delivery systems are being designed to target the nucleus by employing nuclear leader peptides. Even in the absence of any nuclear signal peptides, cationic lipid-based nonviral vector-associated DNA seems to be delivered into the proximity of the cell nucleus, away from the mitochondria in the cytosol. The study on the subcellular localization of all components of a cationic lipid–DNA complex[44] demonstrated that DNA remains in vesicles that were found in a perinuclear location. These results suggest that the lipid–DNA complex was endocytosed and moved toward the nucleus where the endosomes fused and coalesced into large membrane-bound vesicles. Even free, not vector bound, DNA seems to be transported by the cell to the nucleus, as was shown in,[45] where a 28-bp oligonucleotide was injected into the cytoplasm of CF-1 cells and fibroblasts. In all cases, the DNA rapidly and preferentially accumulated in the nucleus.

It is evident that for the successful transport of DNA or DNA–peptide conjugates to mitochondria in living cells, a mitochondria-specific delivery system needs to be developed which must meet the following requirements: (a) it has to bind DNA or DNA–peptide conjugate; (b) it has to neutralize the negative charge of the DNA; (c) it has to protect the DNA and the peptide from proteolytic degradation; (d) it has to mediate and even facilitate the cellular uptake of the complex; (e) it has to transport DNA or DNA–peptide conjugate into close proximity to mitochondria; and (f) it has to selectively release the conjugate upon contact with the mitochondrial membrane. The complexation (binding) of DNA involving protection against hydrolysis and mediation of intracellular uptake has been shown to be accomplished by a whole variety of cationic lipids and polymers, and many of them have already found their way into clinical studies for nuclear–cytosolic gene therapy (see references cited earlier). However, the big challenge of how to transport DNA through the cytosol selectively to the site of mitochondria has neither been properly addressed nor solved.

Mitochondrial Membrane Potential and Mitochondriotropic Compounds: DQAsomes

Since one of the major roles of mitochondria in the metabolism of eukaryotic cells is the synthesis of ATP by oxidative phosphorylation via the respiratory chain, electrons from the hydrogens on NADH and $FADH_2$ are carried along the respiratory chain at the mitochondrial inner membrane, thereby releasing energy that is used to "pump" protons across the inner membrane from the mitochondrial matrix into the intermembrane space. This process creates a transmembrane electrochemical gradient, which includes contributions from both a membrane potential (negative inside) and a pH difference (acidic outside). The membrane potential of mitochondria in vitro is between 180 and 200 mV, which is the maximum a lipid bilayer can sustain while maintaining its integrity.[46] Therefore, positively charged molecules are attracted by mitochondria in response to the highly negative membrane potential. Although most charged molecules cannot enter the mitochondrial matrix because the inner mitochondrial membrane is impermeable to polar molecules, certain amphiphile compounds, however, are able to cross both mitochondrial membranes and accumulate in the mitochondrial matrix in response to the negative membrane potential.

It has long been known that amphiphile compounds with delocalized cationic charge accumulate in mitochondria. Rhodamine 123 is the best known representative of this group used as a stain for mitochondria in living cells.[47] Mitochondrial accumulation of tetraphenylphosphonium chloride and other cationic aryl phosphonium salts was also demonstrated.[48] The mitochondrial membrane potential is considered to be the main driving force for the uptake and retention of lipophilic cations by mitochondria.[49,50] Sufficient lipophilicity combined with delocalization of the positive charge to reduce the free

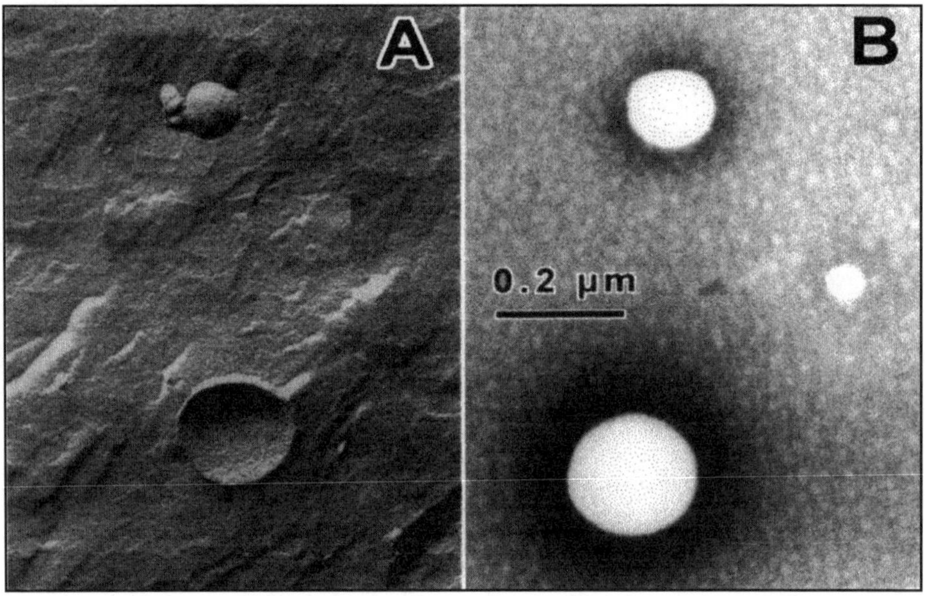

Fig. 4.6. Electron microphotograph of DQAsomes prepared from dequalinium. Panel A – freeze fractured; Panel B – negatively stained. From Ref. (51).

energy change when moving from an aqueous to a hydrophobic environment are believed to be prerequisite for mitochondrial accumulation in response to the mitochondrial membrane potential.[50] It is important that the mitochondrial accumulation and retention of dequalinium (DQA), a single-chain bola amphiphile with two delocalized positive charge centers, was also demonstrated.[50]

Although DQA is widely used in research and pharmaceutical practice, the association properties of this amphiphilic molecule in aqueous medium have rarely been addressed. We have found, however, that upon sonication, DQA forms spherical aggregates with diameters between about 70 and 700 nm.[51,52] It can be clearly seen on the images obtained by electron microscopy (Fig. 4.6). These images strongly indicate the liposome-like aggregation of DQA. Negatively stained samples demonstrate that vesicles (called DQAsomes) are impermeable to the stain. Particle size measurements of DQAsomes stored at room temperature for up to 96 h did not show significant changes in their size distribution as compared with the freshly prepared vesicles measured after 1 h, evidencing that DQAsomes do not precipitate, fuse, or aggregate in solution. The self-assembly behavior of DQA was also confirmed by Monte Carlo simulation.[53] The hypothesis was put forward that, because of their positive charge and mitochondria-tropic properties, DQAsomes can bind or entrap DNA and transport it to mitochondria.

DQAsome–DNA Complexes and Selective Release of DNA by These Complexes at Mitochondria-Like Membranes and Intact Mitochondria

To assess the binding of DNA to DQAsomes we have used the DNA-specific dye SYBR™ Green I. The fluorescent signal of this dye is greatly enhanced when bound to DNA. Non-binding results in loss of fluorescence. We found that increasing the amounts of DQAsomes in a DNA–dye mixture prevents the SYBR™ dye from binding to the DNA, which eventually leads to a complete loss of the fluorescence signal,[51] thus evidencing the strong interaction between DQAsomes and plasmid DNA. It was also shown that the complex formation effectively protects DNA from digestion by nucleases,[54] which is an essential prerequisite for any cell transfection.

In multiple experiments on the cationic carrier-mediated intracellular DNA delivery, it was shown that DNA should both form complex with a carrier and liberate from this complex inside the cell to make the nucleic acid accessible for the transcription machinery. From studies about the mechnism of nuclear–cytosolic transfection using cationic liposomes, it is known that anionic lipids from the cytoplasm-facing monolayer of the cell membrane are able to displace the DNA from its cationic carrier during endocytosis.[15] Opposite to that, our strategy for mitochondria-specific transfection is not to liberate DNA somewhere in the cytoplasm, but rather to transport the DNA into the close proximity to mitochondria followed by the mitochondrial leader peptide-mediated uptake into the mitochondrial matrix. Therefore, DQAsomes must release the DNA (or the DNA–peptide conjugate) at the outer mitochondrial membrane. However, they are not expected to release the DNA upon contact with anionic lipids from the cytoplasmic membrane, i.e. they must remain stably associated with the DNA during endocytosis.

Therefore, we have studied the stability of DNA association with DQAsomes in the presence of anionic liposomes, the lipid composition of which mimicked cytoplasmic (CPM), inner mitochondrial (IMM), and outer mitochondrial membranes (OMM), mitochondrial membranes being enriched with cardiolipin.[55,56] The diameters of CPM-, IMM-, and OMM-liposomes were almost identical (around 80 nm), excluding any impact a different radius of curvature might have on the displacement of the DNA from its carrier. As the result of these experiments we have found that (I) the interaction between anionic liposomes and the DQAsome–DNA complex is very fast (seconds); (II) in the vicinity of a 1:1 charge ratio or even at 1.4-fold excess of anionic charge, DQAsome/DNA complexes do not release any DNA in the presence of CPM liposomes, i.e. liposomes resembling the cytosolic membrane; (III) at similar charge ratios, IMM and OMM liposomes, i.e. liposomes resembling both mitochondrial membranes, are able to displace up to 75% of the DNA from its DQAsomal carrier,[57] see also Fig. 4.7.

The liberation of DNA–peptide conjugates from the cationic vector upon contact with mitochondria is crucial for the subsequent DNA transport into the mitochondrial matrix via the mitochondrial protein import machinery. Therefore, in our further experiments, we have studied the dissociation of DNA from the standard DQAsome–DNA complexes in vitro in the presence of isolated mouse liver mitochondria that have been isolated according to the standard procedure[58] and characterized by electron microscopy, determination of citrate synthese as a marker enzyme for the mitochondrial matrix, and by protein content. To assess the binding of DNA to DQAsomes and the subsequent release upon addition of mitochondria the DNA-specific dye SYBR™ Green I was again employed.

The results of this experiment are shown on Fig. 4.8. Upon rapid addition of mitochondria to DQAsome–DNA suspension, the fluorescence intensity increases following a biphasic pattern. The sudden initial increase of fluorescence (first phase) is most likely caused by the interaction of the dye with mitochondrial endogenous DNA, as we could verify by three control experiments. First, addition of mitochondria to a DNA-free solution of the dye leads to an identical instantaneous fluorescence increase. Second, the treatment of the isolated mitochondria with DNAse prior to adding to the DNA-free dye solution had no impact on fluorescence intensity, ruling out an interaction of SYBR with possibly coisolated contaminating nuclear DNA. Third, the lysis of mitochondria preincubated with SYBR caused a significant increase in the activity of released marker enzyme but not

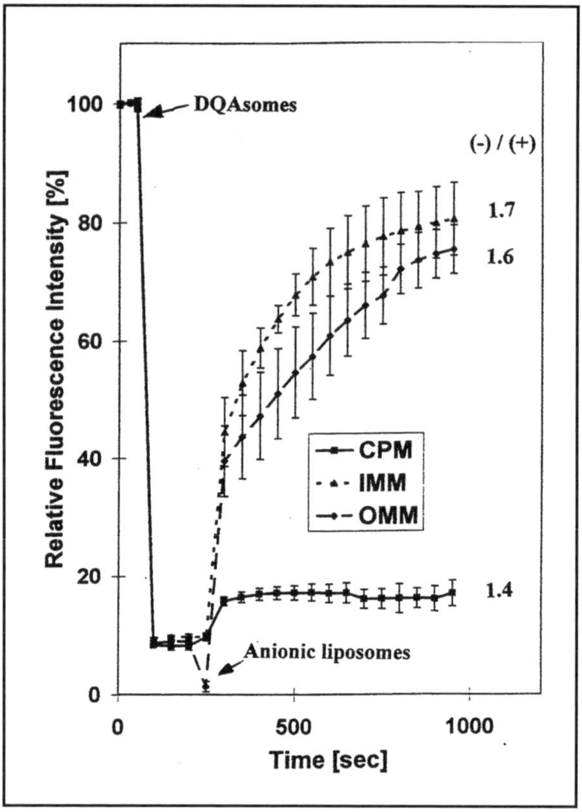

Fig. 4.7. Effect of anionic liposomes on DNA release from DQAsome–DNA complexes. Plasmid DNA (625 ng) was preincubated with SYBR until stabilization of the signal, followed by adding (indicated by arrow) the minimal amount of DQAsome necessary to decrease the signal to the background level. Anionic liposomes were then injected (arrow) at an anionic-to-cationic charge ratio (−)/(+) as shown. The displacement of DNA from its carrier is indicated by the increase of the fluorescence signal. CPM, cytoplasmic membrane-like liposomes; IMM, inner mitochondrial membrane-like liposomes; OMM, outer mitochondrial membrane-like liposomes. All experiments were repeated 5 times. From Ref. (57).

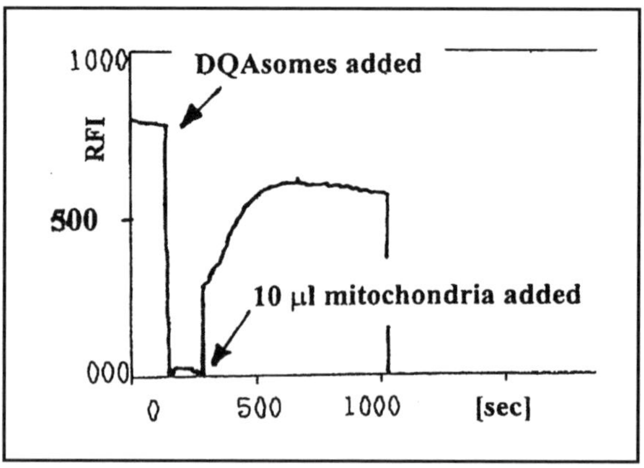

Fig. 4.8. Change of DNA–SYBR™ Green I fluorescence intensity (RFI) due to complexation of DNA by DQAsomes and subsequent liberation of the DNA upon incubation with intact mouse liver mitochondria. From Ref. (58).

in the fluorescence intensity, indicating that all of the endogenous mitochondrial DNA was already associated with the dye prior to the addition of Triton X100. The initial jump in fluorescence intensity (Fig. 4.8) is followed by a slower rise (second phase), which appears to be identical to the time-dependent increase of fluorescence intensity that we have observed during the displacement of DNA from DQAsome–DNA complexes by anionic liposomes.

Here, we have described a whole new strategy for the delivery of DNA into mitochondria. This strategy is based on the self-association of a known mitochondriotropic amphiphilic molecule, dequalinium, into vesicles, called DQAsomes, that meet essential requirements for mitochondrial DNA delivery. Our data strongly suggest that DQAsome–DNA complexes become destabilized upon contact with intact mitochondria, resulting in the release of DNA from DQAsomes. The liberation of DNA from its cationic carrier at the mitochondrial (outer) membrane is a necessary prerequisite for the next step to follow, which is the leader peptide mediated uptake via the mitochondrial protein import machinery. Certainly, this strategy still has to go a long way to provide the functional expression of an exogeneous gene inside the mitochondria. Corresponding experiments are currently in progress.

Literature Cited

1. Roth JA, Swisher SG, Merritt JA, Lawrence DD, Kemp BL, Carrasco CH, El-Naggar AK, Fossella FV, Glisson BS, Hong WK, Khurl FR, Kurie JM, Nesbitt JC, Pisters K, Putnam JB, Schrump DS, Shin DM, Walsh GL. Semin Oncol 1998; 25(3, Suppl. 8):33-37.
2. Lisziewicz J, Sun D, Lisziewicz A, Gallo RC. Gene Ther 1995; 2:218-222.
3. Aalto Setala K, Vuorio E. Ann Med 1997; 29:549-551.
4. Farhood H, Serbina N, Huang L. Biochim Biophys Acta 1995; 1235:289-295.
5. Scheule RK, St George JA, Bagley RG, Marshall J, Kaplan JM, Akita GY, Wang KX, Lee ER, Harris DJ, Jiang C, Yew NS, Smith AE, Cheng SH. Hum Gene Ther 1997; 8:689-707.
6. Filion MC, Phillips NC. Biochim Bhiophys Acta 1997; 1329:345-356.
7. Dokka S, Toledo D, Shi X, CastranovaV, Rojanasakul Y. Pharm Res 2000; 17:521-525.
8. Aberle AM, Tablin F, Zhu J, Walker NJ, Gruenert DC, Nantz MH. Biochemistry 1998; 37:6533-6540.
9. Tang F, Hughes JA. Bioconjug Chem 1999; 10:791-796.
10. Bai HZ, Sawa Y, Zhang WD, Yamakawa T, Morishita R, Kaneda Y, Matsuda H. Ann Thorac Surg 1998; 66:814-819.
11. Dalesandro J, Akimoto H, Gorman CM, McDonald TO, Thomas R, Liggitt HD, Allen MD. J Thorac Cardiovasc Surg 1996; 111:416-421.
12. Yano M, Hiratsuka M, Mora BN, Scheule RK, Patterson GA. Ann Thorac Surg 1999; 68:1810-1814.
13. Boasquevisque CH, Mora BN, Boglione M, Ritter JK, Scheule RK, Yew N, Debruyne L, Qin L, Bromberg JS, Patterson GA. J Thorac Cardiovasc Surg 1999; 117:8-14.
14. Rosenthal FM, Cao L, Tanczos E, Kopp J, Andree C, Stark GB, Mertelsmann R, Kulmburg P. In Vivo 1997; 11:210-208.
15. Xu Y, Szoka FC, Jr. Biochemistry 1996; 35:5616-5623.
16. Khaw BA, Torchilin VP, Vural I, Narula J. Nat Med 1995; 1:1195-1198.
17. Khaw BA, Narula J, Vural I, Torchilin VP. Int J Pharm 1998; 162:71-76.
18. Khaw BA, Vural I, DaSilva J, Torchilin VP. STP Pharma Sci 2000; 10:279-283.
19. Khaw BA, Beller GA, Haber E, Smith TW. J Clin Invest 1976; 58:439-446.
20. Khaw BA, Fallon JT, Beller GA, Haber E. Circulation 1979; 60:1527-1531.
21. Torchilin VP, Ivanov NN, Klibanov AL, Papisov MI, Chebanov SM. In Liposomes as Drug Carriers G Gregoriadis (ed.), Wiley, Avon, UK, 1988, pp. 63-74.
22. Khaw BA, Vural I, Torchilin VP, Haider N, Narula J. In Proc. 23rd International Symposium on Controlled Release of Bioactive Materials, Kyoto, Japan, Controlled Release Society, 1996, pp. 135-136.
23. Sambrook D, Fritsch EF, Maniatis T. Molecular Cloning: A Laboratory Manual, 2nd edn., Cold Spring Harbor Laboratory Press, Cold Spring Harbor Laboratory, New York, 1989.
24. Khaw BA, DaSilva J, Vural I, Narula J, Torchilin VP. J Control Release Submitted.

25. Felgner PL, Ringold GM. Nature 1989; 337:387-388.
26. Prazeres DMF, Schluep T, Cooney C. J Chromatogr 1998; 806:31-45.
27. Khaw BA, Beller GA, Haber E, Smith TW. J Clin Invest 1976; 58:439-446.
28. Khaw BA, Scott J, Fallon JT, Cahill SL, Haber E, Homey C. Science 1982; 217:1050-1053.
29. Torchilin VP, Narula J, Halpern E, Khaw BA. Biochim Biophys Acta 1996; 1279:75-83.
30. Torchilin VP, Trubetskoy VS, Milshteyn AM, Canillo J, Wolf GL, Papisov MI, Bogdanov AA Jr, Narula J, Khaw BA, Omelyanenko VG. J Control Release 1994; 28:45-58.
31. Papa S, Scacco S, Schliebs M, Trappe J, Seibel P. Mol Aspects Med 1996; 17:513-563.
32. Wallace DC. Science 1999; 283:1482-1488.
33. Holt IJ, Harding AE, Morgan-Hughes JA. Nature 1988; 331:717-719.
34. Harding AE. TINS 1991; 14:132-138.
35. Sperl W. Wien Klin Wochenschr 1997; 109:93-99.
36. Seibel P, Trappe J, Villani G, Klopstock T, Papa S, Reichmann H. Nucleic Acids Res 1995; 23:10-17.
37. Chrzanowska-Lightowlers ZM, Lightowlers RN, Turnbull DM. Gene Ther 1995; 2:311-316.
38. Jacobs HT. J Mol Evol 1991; 32:333-339.
39. Kagawa Y, Hayashi JI. Gene Ther 1997; 4:6-10.
40. Collombet JC, Coutelle C. Mol Med Today 1998; 4:31-38.
41. Butow RT, Fox TD. Trends Biochem Sci 1990; 15:465-468.
42. Collombet JM, Wheeler VC, Vogel F. J Biol Chem 1997; 272:5342-5347.
43. Vestweber D, Schatz G. Nature (London) 1989; 338:170-172.
44. Zabner J, Fasbender AJ, Moninger T, Poellinger KA, Welsh MJ. J Biol Chem 1995; 270:18997-19007.
45. Chin DJ, Green GA, Zon G, Szoka FC Jr, Straubinger RM. New Biol 1990; 2:1091-1100.
46. Murphy MP. Biochim Biophys Acta 1989; 977:123-141.
47. Chen LB, Summerhayes IC, Johnson LV, Walsh ML, Bernal SD, Lampidis TJ. Cold Spring Harb Symp Quant Biol 1982; 46:141-155.
48. Rideout D, Bustamante A, Patel J. Int J Cancer 1994; 57:247-253.
49. Lampidis TJ, Hasin Y, Weiss MJ, Chen LB. Biomed Pharmacother 1985; 39:220-226.
50. Weiss MJ, Wong JR, Ha CS, Bleday R, Salem RR, Steele GD Jr, Chen LB. Proc Natl Acad Sci USA 1987; 84:5444-5448.
51. Weissig V, Lasch J, Erdos G, Meyer HW, Rowe TC, Hughes J. Pharm Res 1998; 15:334-337.
52. Weissig V, Torchilin VP. Current Pharm Biotech 2000; 1:325-346.
53. Weissig V, Mögel H-J, Wahab M, Lasch, J. Proceed. Proc Int Symp Control Rel Bioact Mater 1998; 25:312-313.
54. Lasch J. Biol Chem 1999; 380:647-652.
55. Hovius R, Lambrecht H, Nicolay K, de Kruijff B. Biochim Biophys Acta 1990; 1021:217-226.
56. Daum G. Biochim Biophys Acta 1985; 822:1-42.
57. Weissig V, Lizano C, Torchilin VP. Drug Deliv 2000; 7:1-5.
58. Chappell JB, Hansford RG. In Subcellular Components: Preparation and Fractionation, GD Birnie (ed.), Butterworths, London, 1972, pp. 77-91.

Chapter 5

Mannose Receptor-Mediated Gene Delivery into Antigen Presenting Dendritic Cells

Sandra S. Diebold, Christian Plank, Matt Cotten, Ernst Wagner, and Martin Zenke

Abstract

Dendritic cells are professional antigen presenting cells and are unique in their ability to prime naïve T cells. Gene modification of dendritic cells is of particular interest for immunotherapy of diseases where the immune system has failed or is aberrantly regulated, such as in cancer or autoimmune disease, respectively. Dendritic cells abundantly express mannose receptor and mannose receptor-related receptors, and receptor-mediated gene transfer via mannose receptor offers a versatile tool for targeted gene delivery into these cells. Accordingly, mannose polyethylenimine DNA transfer complexes were generated and used for gene delivery into dendritic cells. Mannose receptor belongs to the group of scavenger receptors that allow dendritic cells to take up pathogenic material, which is directed for degradation and MHC class II presentation. Therefore, a limiting step of transgene expression by mannose receptor-mediated gene delivery is endosomal degradation of DNA. Several strategies have been explored to overcome this limitation including the addition of endosomolytic components to DNA transfer complexes like adenovirus particles and influenza peptides. Here, we review the current understanding of mannose receptor-mediated gene delivery into dendritic cells and discuss strategies to identify appropriate endosomolytic agents to improve DNA transfer efficacy.

Introduction

Receptor-Mediated Gene Delivery Is a Powerful Gene Transfer Technique

Gene transfer in medical therapy has to ensure efficient transgene expression in target cells, but additionally has to be safe and readily applicable. Furthermore, the ideal gene delivery systems provide effective and targeted transfer of therapeutic genes into the cell type of choice in vivo. Viral vectors fulfill these criteria to some extent, since they yield

---

Abbreviations: Ad, adenovirus; APC, antigen presenting cells; BSA, bovine serum albumin; DC, dendritic cells; INF, influenza peptide; luc, luciferase; MHC, major histocompatibility complex; Man, mannose; ManR, mannose receptor; PEI, polyethylenimine; pLys, polylysine.

Synthetic DNA Delivery Systems, edited by Dan Luo and W. Mark Saltzman.

efficient transgene expression in vivo and often exhibit cell type specificity. However, viral vectors frequently elicit strong immune responses and thus often can be applied only once. In addition, the widely used retroviral vectors require proliferating cell populations for efficient transduction, and gene transfer efficacy of differentiated, nonproliferating cells is very low. Furthermore, such vectors can induce genetic alterations in transduced cells due to the integration of vector DNA at random sites into the genome. Therefore, nonviral gene delivery systems for in vivo applications are in demand that provide nonimmunogenic transduction of target cells without integration into the host genome. Such delivery systems should be effective also for differentiated cells and additionally should be cell type-specific.

Among the classical nonviral gene delivery techniques, lipofection was used for in vivo gene transfer, but exhibited high systemic side effects and did not provide cell type-specific delivery of the transgene. For selective transfection of specific cell types, transfer complexes have to be targeted to specific surface molecules that are exclusively expressed on these cells. Specific binding to cell surface receptors can be accomplished by including receptor-specific ligands or specific antibodies in DNA transfer complexes.[1] Several receptor-specific gene delivery systems have now been used in vitro and in vivo by targeting, for example, asialoglycoprotein,[2–4] transferrin,[5–7] and mannose receptor.[8–10]

Requirements for Receptor-Mediated Gene Delivery

Receptor-targeted gene delivery into cells involves several steps: 1) DNA condensation and binding to receptor-specific ligand, 2) binding of ligand/DNA complex to cognate receptor, 3) internalization of the receptor/ligand/DNA complex, 4) release of DNA from the endosome, 5) translocation into the nucleus, and 6) transgene expression (Fig. 5.1). Thus, for development of an efficient receptor-based gene delivery system several criteria have to be met. The ligand has to be modified to allow condensation and binding of plasmid DNA, and the polycations polylysine (pLys) or polyethylenimine (PEI), or cationic lipids are frequently used. Following ligand binding to cognate receptor the ligand/DNA transfer complex has to induce internalization of the receptor/ligand/DNA complex and translocation into the endosomal compartment. Thus, only receptors that are internalized upon ligand binding are suitable targets for receptor-mediated gene delivery, and natural receptor ligands or receptor-specific antibodies are employed.[1] After internalization of receptor/ligand/DNA complex, the DNA has to exit the endosomal compartment to escape from degradation. Therefore, transfer complexes often include component(s) with endosomolytic activity. After reaching the cytosol, the DNA has to translocate into the nucleus where it is expressed. This represents a particularly strong barrier since many cell types that are subject to gene modification are fully differentiated and thus postmitotic. In such cells the nuclear membrane stays intact and translocation of DNA into the nucleus relies on active transport mechanisms that so far are poorly understood.

The aim of receptor-mediated gene transfer is to provide a safe and efficient gene delivery system that is cell type-specific. To achieve this goal, the underlying mechanisms of DNA uptake, translocation and expression have to be studied in vitro to address and optimize all the critical steps involved. Such studies are the prerequisite for future applications of cell type-specific DNA transfer complexes in vivo.

Mannose Receptor for Targeted Gene Delivery

Binding of ligand to mannose receptor (ManR) induces the endocytosis of the receptor/ligand complex that is mandatory for a target receptor. At low pH the receptor releases its cargo and recycles back to the cell surface where it can take up and internalize more

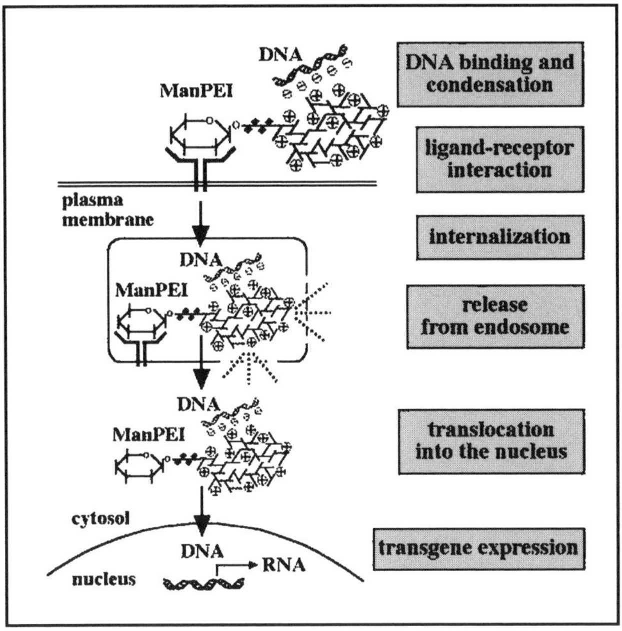

Fig. 5.1. Schematic representation of ManR-mediated gene delivery of ManPEI/DNA transfer complexes. ManPEI conjugates efficiently bind and condense plasmid DNA, and mediate targeted binding to the ManR. Receptor binding induces internalization of the DNA transfer complexes into the endosomal compartment. To avoid degradation of the plasmid DNA in the endosome, release from the endosomal compartment is a critical step in the system. PEI has some endosomolytic activity and has been proposed to cause endosome disruption. The ManPEI/DNA transfer complexes then translocates from the cytoplasm into the nucleus where the transgene is expressed.

ligand in consecutive cycles.[11] ManR is a scavenger receptor that is expressed on antigen presenting macrophages[11,12] and on cultured DC;[13,14] whether DC express ManR in situ remains controversial.[15,16] DC express other members of the multilectin ManR family, like the murine DEC-205 receptor, that has similar binding activities.[16,17] There are, however, functional differences between ManR and DEC-205. DEC-205 was shown to guide the internalized cargo to lysosomes, and not to early endosomes as ManR, and thus to more efficient antigen presentation of internalized antigen.[18] Expression of ManR-related receptors on Langerhans cells has been postulated since these cells take up yeast particles and apparently do not express bona fide ManR.[15,19]

ManR internalizes glycoproteins containing mannose, glucose, fucose and *N*-acetylglucosamine residues that are present in many prokaryotic and fungal cell walls like lipoteichoic acid of Gram-positive or O-linked polysaccharides of Gram-negative bacteria, and in macromolecules of some viruses and parasites. These motifs are absent from mammalian proteins and therefore take part in the self versus nonself discrimination by the immune system. Upon inflammation antigen presenting cells (APC) take up a variety of pathogen-related substances via their scavenger receptors. While macrophages reside at sides of inflammation to participate in the first line of defense against infectious agents by phagocytosing pathogens, DC play a different role. After uptake of pathogenic material, DC process and present antigenic epitopes of pathogenic proteins on their major histocompatibility complex (MHC) class II molecules [see[10] for references]. Upon activation by inflammatory stimuli, DC upregulate their costimulatory properties and migrate to the draining lymph node where they stimulate antigen-specific T cells. By doing so, DC activate the adaptive branch of the immune system, thereby providing the link between innate and adaptive immunity.

Given their unique properties in initiating primary immune responses, several strategies are currently being investigated to employ DC for immunotherapy. Gene transfer into DC was found to be difficult, and adenovirus vectors are so far the most efficient gene delivery system for DC. Adenovirus is, however, highly immunogenic and elicits potent antiviral immune responses that limits its application in medical therapy. Thus, there is a high interest in alternative, preferentially nonviral systems for efficient transduction of DC. ManR-mediated gene transfer represents such a DNA delivery system. ManR and ManR-related receptors are highly expressed on macrophages and DC, and the receptor/ligand complex is internalized upon ligand binding. Accordingly, ManR has been successfully used for receptor targeted DNA delivery into DC.[10,20,21]

A potential limitation of effective gene delivery and transgene expression via ManR is the trapping of plasmid DNA in the endosomal compartment that is specialized for degradation. Therefore, it is important to devise strategies that avoid trapping and degradation of DNA in endosomes, and that ensure effective and rapid release of DNA from this compartment. For disruption of endosomes and DNA release into the cytosol, endosomolytic components are included in DNA transfer complexes. The polycation PEI is assumed to cause endosome swelling and disruption by acting as a "proton sponge".[22,23] Other agents that reduce DNA degradation in endosomes include substances like chloroquine and monensin that prevent acidification of these vesicles.[7,24] Many viruses use receptor-specific internalization pathways for infection of host cells, and consequently have evolved a variety of mechanisms to ensure escape from endosomes. Thus, endosomolytic components in DNA transfer complexes proved to be effective in enhancing transgene expression, include psoralen-inactivated adenovirus particles[25–27] and influenza peptides.[28,29]

In conclusion, the ideal ManR-specific gene deliver system consists of various building blocks including PEI for DNA binding and condensation, mannose residues for receptor-specific binding and endosomolytic components for DNA exit from the endosomal compartment. The impact of the various components for effective gene delivery of DC are discussed.

Current State of Research

ManPEI/DNA Transfer Complexes

The first step for formation of ManPEI/DNA transfer complexes is the binding of plasmid DNA to receptor-specific ligand. Therefore, ligands are modified with the polycations pLys or PEI. pLys is a linear polycation, while PEI is a highly branched network polymer with a high intrinsic proton buffering capacity. pLys and PEI were compared for their transfer efficiency, and PEI was consistently found to be superior to pLys both when unconjugated polycations or transferrin polycation conjugates were used.[20,22] The enhanced transfer efficiency of PEI versus pLys might stem from various properties of this macromolecule that are based on its very high cationic charge density. This results in effective binding of the DNA and in high endosomolytic activity due to "proton sponge" function that leads to endosome swelling and disruption. In addition, PEI was shown to direct translocation of transfer complexes from the cytosol into the nucleus and appears not to interfere with transgene expression.[30] Thus, the use of PEI versus pLys in DNA transfer complexes is expected to have a major impact on transduction efficacy and transgene expression.

For ManR-mediated gene transfer, ManpLys and ManPEI conjugates were employed but these conjugates were never directly compared in one study. ManpLys was used for successful transfection of macrophages in vitro and in vivo[8,9] while ManPEI was shown to be effective for in vitro transfection of BM2 macrophages and primary mouse and human DC.[10,21] Additionally, ManPEI conjugates containing high or low molecular weight PEI of 800 and 50 kDa, respectively, were compared and the molecular weight of the PEI moiety was found to have little influence on transfer efficiency. Furthermore, ManPEI conjugates with a phenylisothiocyanate bridge for linking the mannose residue to PEI were as efficient as ManPEI conjugate with a mannobiose bridge.[10]

Incorporation of Influenza Virus Peptide in ManPEI/DNA Transfer Complexes Increases Transgene Expression

While the linker region of the conjugates and the molecular weight of the PEI moiety did not affect transduction efficiency, agents that influence the release of transfer complexes from the endosomal compartment improved transgene expression significantly. Influenza peptide (INF7),[29] influenza peptide PEI conjugate (INF7PEI), and psoralen-inactivated adenovirus particles[24] were incorporated in ManPEI/DNA transfer complexes. Influenza peptide enhanced transfer efficiencies in a dose-dependent manner with low amounts of INF peptide being most effective (Fig. 5.2A and B). Surprisingly INF7PEI conjugates were found to be ineffective at all concentrations tested (Fig. 5.2A and B; and data not shown). This might be because conjugation of influenza peptide to PEI compromises the endosomolytic activity of the peptide. Highest transfer efficiencies were achieved when

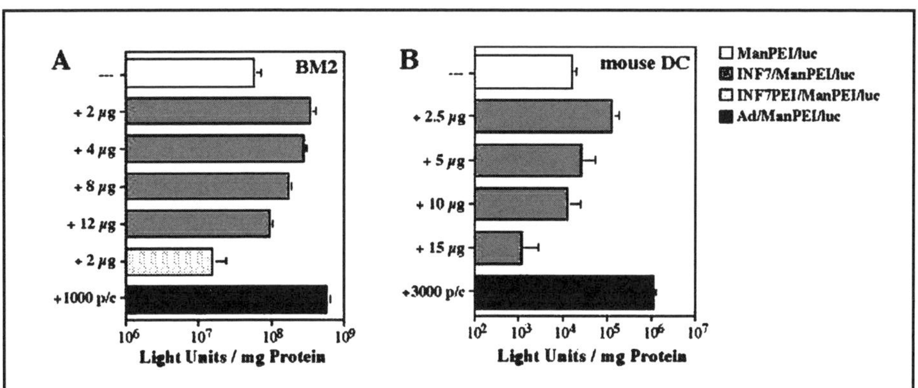

Fig. 5.2. INF7 peptides but not INF7PEI conjugates increase transfer efficiency of ManPEI/DNA transfer complexes. (A) ManPEI/luc transfer complexes were formed in the presence of increasing amounts of INF7 peptide (2–12 µg) or INF7PEI conjugate (2 µg), and BM2 cells were transfected. ManPEI/luc transfer complexes with or without adenovirus particles were used as controls. Luciferase activity was determined 48 h after transfection. A dose-dependent decrease in luciferase activity was observed, with 2 µg of INF7 peptide being most efficient; lower concentrations of INF7 peptide yielded lower transgene expression (data not shown). ManPEI/DNA transfer complexes containing INF7PEI conjugate did not increase luciferase activity at all concentrations tested (data not shown). (B) ManPEI/DNA transfer complexes with INF7 peptide (2.5–15 µg) were generated as in (A) and transfected into mouse DC. Again low amounts INF7 increased transgene expression that was however 10-fold lower than transgene expression from Ad/ManPEI/DNA transfer complexes.

psoralen-inactivated adenovirus particles were employed which was observed for BM2 cells and was especially prominent for DC (Fig. 5.2A and B). These data indicate that for DC one of the limiting steps for efficient transgene expression is exit of plasmid DNA from the endosomal compartment rather than cell entry via the cell surface receptor.

ManPEI/DNA Complex Formation Requires Specific Salt Concentration

The size of DNA transfer complex represents another important parameter for efficient cellular uptake and transgene expression. DNA transfer complexes that form large aggregates are not (or very poorly) taken up via the receptor internalization route. The size of DNA transfer complex is influenced by the ionic strength of the buffer when the transfer complex is formed.[31] Compaction of the DNA into spherical particles is also a critical parameter for DNA trafficking to the nucleus.[30] When Man/PEI/DNA transfer complexes were formed in aqueous solution without salt, in 20 mM HEPES or in HEPES buffered saline (HBS), best transfer efficiencies were consistently found for HBS (Fig. 5.3A

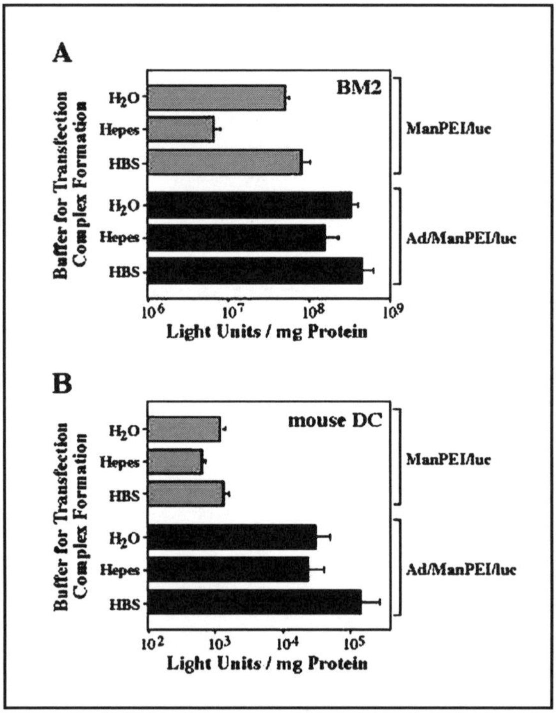

Fig. 5.3. Salt concentration influences the formation of transfer complexes and transfer efficiency. (A) ManPEI/DNA transfer complexes with luciferase reporter plasmid were formed with and without adenovirus particles in H_2O, HEPES (20 mM, pH 7.5), and HEPES buffered saline (HBS, 20 mM HEPES, pH 7.5 plus 150 mM NaCl), respectively. BM2 cells were transfected and cell lysates were tested for luciferase activity 48 h after transfection. For ManPEI/luc and Ad/ManPEI/luc transfer complexes transgene expression was strongest when HBS was used for complex formation. (B) ManPEI/DNA transfer complexes with luciferase reporter plasmid were generated as in (A) and transfected into mouse DC. The overall transgene activity achieved was lower than for BM2 cells but again transfer complexes formed in HBS were most efficient.

and B). This result was obtained for BM2 macrophages and primary mouse DC, and was independent of the linker region of the ManPEI conjugate, the molecular weight of the PEI moiety, and the presence of psoralen-inactivated adenovirus particles in transfer complexes (Fig. 5.3A and B; and data not shown). Interestingly, ManPEI/DNA transfer complexes generated in water were only marginally less efficient than those formed in HBS, while the use of 20 mM HEPES buffer for complex formation led to a much more prominent decline in transgene expression. This result suggests that compaction of plasmid DNA by ManPEI conjugates is influenced by salt concentration during complex formation.

Uptake of ManPEI/DNA Transfer Complexes Is Receptor-Specific

The most prominent advantage of receptor-mediated gene delivery over other transfection systems is that DNA transfer complexes are targeted to specific cell types. Binding and internalization of ManPEI/DNA transfer complexes to ManR is inhibited by excess of ligand (Fig. 5.4B and C). When mannosylated bovine serum albumin (ManBSA) was added prior to ManPEI/DNA transfer complexes, transgene expression of the luciferase reporter plasmid (luc) was reduced by 10,000-fold in BM2 cells, while unmodified BSA left luciferase activity unaffected (Fig. 5.4B). Essentially the same result was obtained for mouse DC, thereby demonstrating the ManR-specific uptake of ManPEI/DNA transfer complexes.[10]

Adenovirus particles are frequently incorporated in PEI/DNA transfer complexes (in the following referred to as Ad/ManPEI/DNA transfer complexes) and bind to PEI via

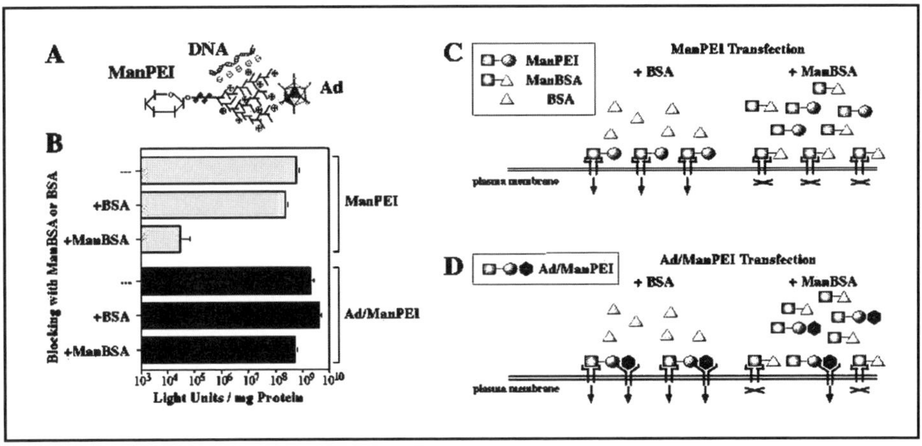

Fig. 5.4. ManPEI/DNA but not Ad/ManPEI/DNA transfer complexes are taken up via ManR. (A) Schematic representation of ManPEI/DNA transfer complexes containing adenovirus particles. The adenovirus particles bind to the polycation via negative charges on the viral capsid. (B) ManPEI/DNA transfer complexes efficiently transduce BM2 cells as do Ad/ManPEI/DNA transfer complexes. Treatment of cells with ManBSA reduces transfer efficiency of ManPEI/DNA but not that of Ad/ManPEI/DNA transfer complexes, while BSA leaves transfer efficiency unaffected. This indicates that ManPEI/DNA but not Ad/ManPEI/DNA transfer complexes enter cells by a ManR-specific uptake mechanism. (C) Schematic representation of ManPEI transfection where receptor specific binding of transfer complexes is blocked by addition of an excess of ManBSA but not BSA, demonstrating that ManPEI/DNA transfer complexes are taken up via ManR. (D) Schematic representation of Ad/ManPEI transfection that is not blocked by ManBSA or BSA, indicating that these transfer complexes are taken up by an alternative pathway (probably via the adenovirus infection route).

charged interactions with virus capsid proteins (Fig. 5.4A). At low adenovirus particle numbers in Ad/ManPEI/DNA transfer complexes, transduction efficiency is only slightly elevated, while at high adenovirus particle numbers cell viability is compromised.[10] In the presence of psoralen-inactivated adenovirus particles at optimal particle number, uptake of Ad/ManPEI/luc transfer complexes and expression of luciferase reporter is only marginally inhibited by addition of excess ManBSA in BM2 cells (Fig. 5.4B). This indicates that Ad/ManPEI/DNA transfer complexes are no longer exclusively taken up via the ManR-specific internalization pathway and adenovirus capsid itself appears to act as ligand for uptake via the adenovirus infection route (Fig. 5.4D). In BM2 cells incorporation of adenovirus particles in ManPEI/DNA transfer complexes leads only to a 10-fold increase in transgene expression (Figs. 5.2A and 4B), and thus it appears that in BM2 cells the ManR and adenovirus internalization routes are equally efficient. This is different for DC where the presence of adenovirus particles in ManPEI/DNA transfer complexes enhances transgene expression by at least 100-fold (Fig. 5.2B). In DC, even in the presence of low adenovirus particle numbers in Ad/ManPEI/DNA transfer complex, uptake is no longer exclusively occurring via ManR and consequently is no longer blocked by ManBSA.[10] This indicates that in this cell type uptake via the adenovirus infection pathway is much more efficient than uptake via ManR.

Discussion

Receptor-mediated gene delivery includes a number of critical steps that affect transfer efficiency (Fig. 5.1). It is therefore important to identify the limiting step(s) for efficient transfection of DC. DC used in these studies expressed high levels of ManR as shown by staining with a ManR-specific antibody and FITC-labeled ManBSA.[10] Therefore, binding of ManPEI/DNA transfer complexes to ManR appears not to represent the limiting step. From the data obtained so far, it is most likely that endosomal degradation of the endocytosed plasmid DNA severely compromises transgene expression. Further indications towards endosomal trapping as the limiting step stems from our observation that INF7 peptide significantly enhances transgene expression in DC.

Psoralen-inactivated adenovirus particles act as a potent endosomolytic agent, for example, in transferrinfection that employs transferrin pLys/DNA transfer complexes.[24] In transferrinfection adenovirus particles in transfer complexes were shown not to alter the specificity of receptor-mediated internalization.[25,27] However, for transfection of DC by Ad/ManPEI/DNA transfer complexes, ManR-specific uptake was not the exclusive internalization pathway.[10] This finding is in accord with the observation that Ad/PEI/DNA transfer complexes without Man residues are as effective as Ad/ManPEI/DNA transfer complexes in gene delivery into DC.[21] Ad/PEI transfection is in principle suitable for all cell types that are susceptible to adenovirus infection and thus the major advantage of receptor-mediated gene delivery over other gene transfer systems, receptor and cell type specificity, is lost. Therefore, adenovirus particles despite their potent endosomolytic activity are not suitable as a component of the ManPEI transfer system. Thus, incorporation of other agents that induce efficient DNA exit from the endosomal compartment and improve the ManPEI gene delivery system are in demand. One such potential agent might be the pore-forming protein listeriolysin O (LLO) of the bacteria *Lysteria monocytogenes*:[32] LLO allows the bacteria to escape from the host cell vacuole without damaging the cytoplasma membrane and thus represents an endosomolytic agent with an activity that is restricted to intracellular membranes in particular of the endosomal compartment.

In vivo gene transfer via ManR was so far only tested for i.v. injection of ManpLys[9] and mannosylated cationic liposomes.[33] In both studies transgene expression was mainly found in lung, liver, and spleen. Transgene expression in the liver was detected in non-parenchymal cells[33] and in the spleen in small cells in the white pulp, marginal zones, and subcapsular region.[9] It was concluded that these transduced cells represent liver Kuppfer cells or splenic macrophages, but there is so far no direct evidence that transgene expression was restricted to these cells and unspecific transduction of other cell types can not be excluded. Administration by i.v. injections is appropriate to study some aspects of uptake and transgene expression by splenic DC, but more interesting would be s.c. injection of transfer complexes to investigate transgene expressing DC in draining lymph nodes. The latter administration route mimics an infection in the periphery when peripheral DC become activated and migrate to the draining lymph nodes to initiate adaptive immune responses. Such studies are of fundamental importance to further develop ManR-mediated gene transfer for DC-based approaches in immunogene therapy. In addition, expression of ManR and ManR-like receptors is not exclusively restricted to APC. ManR was found to be expressed on some endothelial cell types, perivascular microglia and mesangial cells,[15,34] and so far it remains unclear how transduction of these cell types by in vivo gene transfer affects the therapeutical outcome. The use of DC-specific promoters that drive transgene expression might provide a way to overcome these limitations.

Taken together, in vivo gene transfer by ManR-mediated gene delivery is poorly studied so far and there is high demand for experiments that explicitly address questions about the cell types transduced. In addition, such in vivo experiments have to demonstrate that transgene expression levels can be achieved that are high enough to elicit potent therapeutic effects.

Acknowledgments

We thank Steffen Kurz for recombinant GM-CSF, Margaretha Kursa and Irene Gallagher for expert technical and secretarial assistance, respectively, and Caetano Reis e Sousa for critical reading of the manuscript. This work was in part supported by a grant from the Deutsche Forschungsgemeinschaft (DFG, SFB 506) to M. Z. and a grant from the MDC Gene Therapy Program to S. D.

Literature Cited

1. Kircheis R, Kichler A, Wallner G, Kursa M, Ogris M, Felzmamr T, Buchberger M, Wagner E. Gene Ther 1997; 4:409-418.
2. Wu GY, Wu CH. J Biol Chem 1987; 262:4429-4432.
3. Wu GY, Wu CH. J Biol Chem 1988; 263:14621-14624.
4. Wu CH, Wilson JM, Wu GY. J Biol Chem 1989; 264:16985-16987.
5. Cotten M, Langle-Rouault F, Kirlappos H, Wagner E, Mechtler K, Zenke M, Beug H, Birnstiel ML. Proc Natl Acad Sci USA 1990; 87:4033-4037.
6. Wagner E, Zenke M, Cotten M, Beug H, Birnstiel ML. Proc Natl Acad Sci USA 1990; 87:3410-3414.
7. Zenke M, Steinlein P, Wagner E, Cotten M, Beug H, Birnstiel ML. Proc Natl Acad Sci USA 1990; 87:3655-3659.
8. Erbacher P, Bousser MT, Raimond J, Monsigny M, Midoux P, Roche AC. Hum Gene Ther 1996; 7:721-729.
9. Ferkol T, Perales JC, Mularo F, Hanson RW. Proc Natl Acad Sci USA 1996; 93:101-105.
10. Diebold SS, Kursa M, Wagner E, Cotten M, Zenke M. J Biol Chem 1999; 274:19087-19094.
11. Stahl P, Schlesinger PH, Sigardson E, Rodman JS, Lee YC. Cell 1980; 19:207-215.
12. Stahl P, Gordon S. J Cell Biol 1982; 93:49-56.
13. Sallusto F, Cella M, Danieli C, Lanzavecchia A. J Exp Med 1995; 182:389-400.

14. Avrameas A, McIlroy D, Hosmalin A, Autran B, Debre P, Monsigny M, Roche AC, Midoux P. Eur J Immunol 1996; 26:394-400.
15. Linehan SA, Martinez-Pomares L, Stahl PD, Gordon S. J Exp Med 1999; 189:1961-1972.
16. Kato M, Neil TK, Fearnley DB, McLellan AD, Vuckovic S, Hart DN. Int Immunol 2000; 12:1511-1519.
17. Jiang W, Swiggard WJ, Heufler C, Peng M, Mirza A, Steinman RM, Nussenzweig MC. Nature 1995; 375:151-155.
18. Mahnke K, Guo M, Lee S, Sepulveda H, Swain SL, Nussenzweig M, Steinman RM. J Cell Biol 2000; 151:673-684.
19. Reis e Sousa C, Stahl PD, Austyn JM. J Exp Med 1993; 178:509-519.
20. Diebold SS, Cotten M, Wagner E, Zenke M. Adv Exp Med Biol 1998; 451:449-455.
21. Diebold SS, Lehrmann H, Kursa M, Wagner E, Cotten M, Zenke M. Hum Gene Ther 1999; 10:775-786.
22. Boussif O, Lezoualc'h F, Zanta MA, Mergny MD, Scherman D, Demeneix B, Behr JP. Proc Natl Acad Sci USA 1995; 92:7297-7301.
23. Godbey WT, Wu KK, Mikos AG. Proc Natl Acad Sci USA 1999; 96:5177-5181.
24. Cotten M, Saltik M, Kursa M, Wagner E, Maass G, Birnstiel ML. Virology 1994; 205:254-261.
25. Curiel DT, Agarwal S, Wagner E, Cotten M. Proc Natl Acad Sci USA 1991; 88:8850-8854.
26. Cotten M, Wagner E, Zatloukal K, Phillips S, Curiel DT, Birnstiel ML. Proc Natl Acad Sci USA 1992; 89:6094-6098.
27. Wagner E, Zatloukal K, Cotten M, Kirlappos H, Mechtler K, Curiel DT, Birnstiel ML. Proc Natl Acad Sci USA 1992; 89:6099-6103.
28. Wagner E, Plank C, Zatloukal K, Cotten M, Birnstiel ML. Proc Natl Acad Sci USA 1992; 89:7934-7938.
29. Plank C, Oberhauser B, Mechtler K, Koch C, Wagner E. J Biol Chem 1994; 269:12918-12924.
30. Pollard H, Remy JS, Loussouarn G, Demolombe S, Behr JP, Escande D. J Biol Chem 1998; 273:7507-7511.
31. Perales JC, Ferkol T, Molas M, Hanson RW. Eur J Biochem 1994; 226:255-266.
32. Decatur AL, Portnoy DA. Science 2000; 290:992-995.
33. Kawakami S, Sato A, Nishikawa M, Yamashita F, Hashida M. Gene Ther 2000; 7:292-299.
34. Groger M, Holnthoner W, Maurer D, Lechleitner S, Wolff K, Mayr BB, Lubitz W, Petzelbauer P. J Immunol 2000; 165:5428-5434.

CHAPTER 6

Electro-Gene-Transfer: A New Approach for Muscle Gene Delivery

Elena Fattori, Nicola La Monica, Gennaro Ciliberto, and Carlo Toniatti

Abstract

Gene transfer into skeletal muscle cells by direct injection of naked plasmid DNA results in sustained gene expression. Intramuscular injection of plasmid DNA might thus be used to correct myopathies, to secrete locally or systematic therapeutic proteins and to elicit an immune response against specific antigens. However, the potential utility of this technique for gene application in humans is limited by the poor transduction efficiency and the low and highly variable level of gene expression. Different methods are thus being developed to increase the efficiency of gene transfer in muscles. It has been recently reported that a dramatic improvement of DNA transfer is achieved by applying an electric field to the muscle fibers subsequent to local DNA injection. Electro-gene-transfer increases gene expression by several orders of magnitude and strongly reduces interindividual variability. Electroinjection of genes encoding for secreted proteins resulted in sustained expression and disease correction in animal models of gene therapy. Moreover, the immunogenicity of DNA vaccines is dramatically increased when antigen-encoding plasmids are delivered by this technique. This technique may thus have broad and important applications in human gene therapy. This review provides a brief overview of the theory of electro-gene-transfer and describes parameters governing its efficiency in muscle. We also summarize the results obtained with electro-gene-transfer in animal models to date and the technical issues that must be solved before its use for human therapy can be considered.

Direct Intramuscular Injection of Naked Plasmid DNA

Intramuscularly injected naked plasmid DNA is internalized and expressed by skeletal muscle cells of rodents and primates.[1,2] Plasmid DNA is taken up by myofibers adjacent to the site of injection and is maintained and expressed up to 19 months after injection.[2] This long-term stability and transcriptional activity is made possible because myofibers are

Abbreviations: DMD, Duchenne muscular dystrophy; PVP, polyvinyl pyrrolidone; PVA, polyvinyl alcohol; IL-5, interleukin-5; SEAP, secreted alkaline phosphatase; Epo, erythropoietin; VEGF, vascular endothelial growth factor.

Synthetic DNA Delivery Systems, edited by Dan Luo and W. Mark Saltzman.

terminally differentiated and long-lived cells that cannot reenter mitosis, and therefore do not dilute out the internalized plasmids.[3]

Direct intramuscular injection of plasmid DNA has several advantages over virus-based delivery systems. It is a straightforward technique, which can be easily applied, and allows for repeated administration of plasmids with no decrease in efficacy.[2,4] Moreover, plasmids are simple to construct, do not raise the safety concerns associated with the use of viral vectors, and can be prepared in large amounts by using relatively simple technologies.[5] Therefore, muscle DNA injection has great potential for the treatment of several inherited and acquired human diseases.[6] Gene delivery into skeletal muscle of the appropriate genes could be used to cure single-gene recessive inherited diseases primarily affecting the muscle fibers, such as DMD.[7] To use the muscle as a bioreactor for producing proteins that exert a therapeutic effect is also a possibility. Indeed, the muscle tissue offers several advantages in this direction: it has a rich blood supply, constitutes over 40% of the entire body mass, has a great capacity for protein synthesis and secretion, and is readily accessible for direct injection.[6] Therefore, systemic diseases requiring repeated administrations of proteins, such as hemophilia, growth hormone deficiency, anemia, interferon-sensitive viral hepatitis, or some lysosomal storage diseases, could in theory be treated by infrequent injection of plasmid DNA expressing the desired therapeutic protein.[3,6] Moreover, DNA muscle injection in animal models has proven capable of inducing humoral and cell-mediated immunity against a wide range of viral, bacterial, parasitic, and tumor-specific antigens and thus has been proposed as a potential approach to genetic vaccination.[8,9]

In spite of this positive evidence, the progression of intramuscular gene delivery towards therapeutic applications has been greatly hampered by the poor efficiency of gene transfer upon intramuscular injection. In fact, the majority of DNA is degraded or removed before entering muscle cells and no more than 1–5% of the myofibers surrounding the injection site take up and express DNA.[10] In addition, this technique does not guarantee reproducible results, and highly variable levels of gene expression are usually observed in animals.[1,2,10] Another limitation is that the efficiency of gene transfer progressively decreases in proportion to the size of animals.[11–13] The reasons for this are not clear, but the size and the length of myofibers from different species should not be a major cause for this drop in transduction efficiency: myofibers from different species are similar in diameter, and expression is similar in different muscles within a species, despite the fact that the length of myofibers can be very different.[12] A possible explanation comes from comparative studies performed in mice and nonhuman primates which suggest that the poor level of transduction in monkey muscles might be due to the lower number of nuclei per muscle area and therefore to less plasmid DNA being available for active gene expression.[12] Alternatively, the connective tissue, which is more abundant in the perimysium of primates than in rodents, might impede distribution of the plasmid DNA within the tissue and consequently reduce uptake by cells.[12] Moreover, recent evidence suggests that the hydrostatic pressure created by injection plays a role in facilitating DNA intramuscular distribution and uptake in rodents.[14] The lower transduction efficiency in larger animals might thus be due to the injection volumes being much smaller in proportion than those used in mice, leading to less hydrostatic pressure.

In conclusion, despite its great potential for a wide range of therapeutic applications, the clinical utility of straight naked DNA injection is currently limited and confined to its use for generating DNA-based vaccine, in which localized and low-efficiency gene delivery may be sufficient to generate a long-term, albeit weak, immune response also in primates.[15–18]

for number, magnitude, and duration of pulses. The longevity of expression might instead differ according to treatment, although other factors such as the nature of the delivered transgene, the architecture and quality of the plasmids, and the strain of mice used most probably play a role too. A precise and parallel comparison of the different electrical treatments using the same plasmids and mouse strains would thus be required to clarify these points.

The DNA electroinjection technique is still in its infancy and only a limited number of studies have been performed to study the mechanisms involved. Three different studies have quantitatively investigated the influence of different electrical parameters on the efficiency of electroporation and provided insight into the physical mechanisms of this phenomenon.[39,40,52] These studies were performed in mice and 51Cr-EDTA was used as a quantitative marker to measure uptake by muscle tissues subjected to different types of electric field through various electrode configurations. 51Cr-EDTA is a hydrophilic molecule originally used to examine renal clearance.[53] 51Cr-EDTA distributes freely in the vascular extracellular compartments, but does not enter the intracellular compartment unless access is provided, e.g. by electroporation. Because of its rapid renal clearance, the uptake by muscles can be measured shortly after the administration of the radioactive compound. Using this approach, it has been demonstrated that the threshold electric field for efficient electroporation is dependent on pulse length and number.[39] DNA transfer is optimal for field strength just above the permeabilization threshold, as measured by 51Cr-EDTA uptake, thus demonstrating that membrane electroporation is critical for optimal DNA entry.[40,52]

Using the same type of experimental setting, Bureau and colleagues evaluated the effects of various combinations of square wave pulses of different strength and duration on cell permeabilization and DNA electroporation in mice muscles. Interestingly, 1 pulse of 800 V/cm and 100 μs duration (high-voltage short-pulse treatment) or a series of 4 pulses of 80 V/cm and 83 ms duration (low-voltage, short-pulse treatment) did not increase muscle permeability, as measured by 51Cr-EDTA uptake, and only slightly augmented efficacy of DNA transduction as compared to naked DNA injection. By contrast, 1 high-voltage short-pulse followed by 4 low-voltage, short-pulses induced muscle permeabilization and enhanced gene transfer.[52] This confirms the tight association between permeabilization and electrophoretic forces for intramuscular DNA electrotransfer. In agreement with the theoretical models of electroporation, the uptake of small molecules was increased when they were injected before or after the electrical stimulation, while DNA had to be injected before the application of the electric pulse.[28,34,37] This indicates that electric field is needed as a direct driving force for DNA entry into myofibers.[28,29] Efficient electroporation can thus be obtained by a combination of pulses of sufficient electric field and duration to obtain both membrane permeabilization and DNA translocation inside the cell.[52] On the other hand, an overly intense electric field or overlong pulses can cause tissue damage, reduce gene expression, and in certain instances reduce its longevity.[28,29,40,52]

Perspectives

Electroporation experiments in mice suggest that a large injection volume is preferable but not indispensable to achieve an effective transduction of myofibers. In fact, significant expression levels of reporter gene were measured upon inelectroinjection of a plasmid diluted in a 5 μl volume, which proved almost ineffective in the case of simple injection.[14] However, the quantity of DNA that must be electroinjected to achieve efficient expression in larger animals is greater than expected from a linear extrapolation of the quantities used in rodents.[28,54] It is worth noting that with larger animals, hundreds of micrograms or milligrams of plasmid DNA were electroinjected into a single site.[28,54] It is thus possible that delivering DNA by multiple electroinjections in different muscles could significantly increase efficiency in larger animals. In connection to this latter point, the precise identification of those muscles more susceptible to electroporation would also prove useful.[28,43]

At this stage, scaling-up electroinjection technology is still an unresolved issue. Moreover, it is questionable whether the electrical parameters used so far in animal models could be tolerated by humans: it is probable that they will need to be further modified to minimize pain and reduce tissue damage. In conclusion, although muscle electroporation still represents the most efficient nonviral system for delivering genes to skeletal muscle, it must be extensively ameliorated before its use for human therapy can be considered.

Acknowledgment

We thank J. Clench for editing the manuscript.

References

1. Wolff JA, Malone RW, Williams P et al. Science 1990; 247:1465-1468.
2. Wolff JA, Ludtke JJ, Acsadi, G et al. Hum Mol Genet 1992; 1:363-369.
3. Van Deutekom JCT, Hoffman EP, Hueard, J. Mol Med Today 1998; 4:214-220.
4. Rizzuto G, Cappelletti M, Maione D et al. Proc Natl Acad Sci USA 1999; 96:6417-6422.
5. Horn AN, Meek JA, Budahazi G et al. Hum Gene Ther 1995; 6:565-573.
6. Svensson EC, Tripathy SK, Leiden JM. Mol Med Today 1996; 2:166-172.
7. Engel AG, Yamamoto M, Fischbeck KH. In Myology, AG Engel, C Franzini-Armstrong (eds.), McGraw-Hill, New York, 1994, pp. 1133-1187.
8. Nabel G, Felgner PL. Trends Biotechnol 1993; 11:211-215.
9. Cohen AD, Boyer JD, Weiner DB. FASEB J 1998; 12:1611-1626.
10. Doh SG, Vahlsing HL, Hartikka J et al. Gene Ther 1997; 4:648-663.
11. Kitsis RN, Buttrick PM, McNally EM et al. Proc Natl Acad Sci USA 1991; 88:4138-4142.
12. Jiao S, Williams P, Berg RP et al. Hum Gene Ther 1992; 3:21-33.
13. Vitadello M, Schiaffino WV, Picard A et al. Hum Gene Ther 1994; 5:11-18.
14. Dupuis M, Denis-Mize K, Woo C et al. J Immunol 2000; 165:2850-2858.
15. Liu M, McClements W, Ulmer J et al. Vaccine 1999; 17:909-912.
16. Wang R, Doolan DL, Le TP et al. Science 1998; 282:476-480.
17. MacGregor RR, Boyer JD, Ugen KE, Lacy KE, Gluckman SJ, Bagarazzi ML, Chattergoon MA, Bainè Y, Higgins TJ, Ciccarelli, RB, Coney LR, Ginsberg RS, Weiner DB. J Infect Dis 1998; 178:92-100.
18. Calarota S, Bratt G, Nordlund S et al. Lancet 1998; 351:1320-1325.
19. Alila H, Coleman M, Nitta H et al. Hum Gene Ther 1997; 8:1785-1795.
20. Hartikka J, Sawdey M, Cornefert-Jensen F et al. Hum Gene Ther 1996; 7:1205-1217.
21. Baranov A, Glazkov P, Kiselev A et al. Gene Ther 1999; 6:1406-1414.
22. Li X, Eastman EM, Schwarts RJ et al. Nat Biotech 1999; 17:241-245.
23. Mumper RJ, Duguid JG, Khursheed A et al. Pharm Res 1996; 13:701-709.
24. Anwer K, Earle KA, Shi M et al. Pharm Res 1999; 16:889-895.
25. Lemieux P, Guerin N, Paradis G e al. Gene Ther 2000; 7:986-991.
26. Singh M, Briones M, Ott G et al. Proc Natl Acad Sci USA 2000; 97:811-816.
27. Aihara H, Miyazaki J. Nat Biotech 1998; 16:867-870.
28. Mir LM, Bureau MF, Gehl J e al. Proc Natl Acad Sci USA 1999; 96:4262-4267.
29. Mathiesen I. Gene Ther 1999; 6:508-514.
30. Neumann E, Schaefer-Ridder M, Wang Y et al. EMBO J 1982; 1:841-845.
31. Chu G, Hayakawa H, Berg P. Nucleic Acids Res 1987; 15:1311-1326.
32. Andreason GL, Evans GA. Biotechniques 1988; 6:650-660.
33. Weaver JC, Chizmadzhev YA. Biolectrochem Bioenerg 1996; 41:135-160.
34. Neumann E, Kakorin S, Toesing K. Biolectrochem Bioenerg 1999; 48:3-16.
35. Jaroszeski MJ, Gilbert R, Nicolau C et al. Adv Drug Deliv Rev 1999; 35:131-137.

36. Teissie J, Rols MP. In Charge and Field Effects in Biosystems-3, MJ Allen, SF Cleary, AE Sowers, D Shillady (eds.), Birkhauser Press, Boston, pp. 285-301.
37. Lurquin PF. Mol Biotechnol 1997; 7:5-35.
38. Lee RC, Canady DJ, Hammer SM. J Burn Care Rehabil 1993; 5:529-539.
39. Gehl J, Mir L. Biochem Biophys Res Comm 1999; 261:377-380.
40. Gehl J, Thyge H, Sorensen H, et al. Biochim Biophys Acta 1999; 1428:233-240.
41. Vicat JM, Boisseau S, Jourdes P, Lane H, Wion D, Bouali-Benazzouz R, Benadid AL, Berger F. Hum Gene Ther 2000; 11:909-916.
42. Bettan M, Emmanuel F, Darteil R et al. Mol Ther 2000; 2:204-210.
43. Rizzuto G, Cappelletti M, Mennuni C et al. Hum Gene Ther 2000; 11:1891-1900.
44. Kreiss P, Bettan M, Crouzet J et al. J Gene Med 1999 1:245-250.
45. Maruyama H, Sugawa M, Moriguchi Y et al. Hum Gene Ther 2000; 11:429-437.
46. Abruzzese RV, Godin D, Mehta V et al. Mol Ther 2000; 2:276-287.
47. Abruzzese RV, Godin D, Burcin M et al. Hum Gene Ther 1999; 10:1499-1507.
48. Widera G, Austin M, Rabussay D et al. J Immunol 2000; 164:4635-4640.
49. Martin T, Parker SE, Hedstrom R et al. Hum Gene Ther 1999; 10:759-768.
50. Kadowaki S, Chen Z, Asanuma H et al. Vaccine 2000; 18:2779-2788.
51. Zucchelli S, Capone S, Fattori E et al. J Virol 2000; 74:11598-11607.
52. Bureau MF, Gehl J, Deleuze V et al. Biochim Biophys Acta 2000; 1474:353-359.
53. Stacy BD, Thorburn GD. Science 1996; 152:1076-1077.
54. Draghia-Aki R, Fiorotto ML, Anne Hill L et al. Nat Biotech 1999; 17:1179-1183.

CHAPTER 7

Tumor-Targeted Gene Transfer with DNA Polyplexes

Manfred Ogris and Ernst Wagner

Abstract

Systemic gene delivery systems are needed for therapeutic applications; in some situations, target cells might be spread throughout the organism, as in the case of cancer metastases, which can be reached only via the systemic route. Within the class of nonviral vectors, polymer-based transfection particles named DNA polyplexes and lipid-based systems named DNA lipoplexes are being developed for this purpose. For systemic circulation, masking the surface charge of DNA complexes has to be accomplished to avoid interactions with plasma components, erythrocytes, and the reticuloendothelial system. Among other vector formulations, polyplexes based on polyethylenimine (PEI), shielded with polyethylene glycol (PEG), and linked to the receptor binding ligands transferrin (Tf) or epidermal growth factor (EGF) have been developed. Complexes were found to mediate efficient gene transfer into tumor cell lines in a receptor-dependent and cell-cycle-dependent manner. Systemic administration of surface-shielded Tf-PEI polyplexes into the tail vein of mice resulted in preferential gene delivery into distantly growing subcutaneous tumors. In contrast, application of positively charged PEI polyplexes directed gene transfer primarily to the lung.

Introduction

Gene therapy approaches seem to be particularly attractive for applications where the therapeutic agent should be directed only to the site of disease. In theory multiple levels of specificity are attainable by exploiting specific delivery mechanisms to the target cells (biochemical or physical targeting), intracellular events (e.g. targeting of replicating cells only), a controlled tissue-specific expression by the appropriate promoter/enhancer elements (transcriptional targeting), and utilizing biological amplification mechanisms (DNA into many copies of RNA into many copies of protein) within the target cells. However, the lack of effective and targeted viral or nonviral vectors is a major bottleneck for progress in somatic gene therapy. Major limitations for nonviral gene transfer particles[1] include the degradation of the DNA in the biological environment, unspecific sequestration

Abbreviations: EGF, epidermal growth factor; HBS, HEPES buffered saline; PEG, polyethylene glycol; PEI, polyethylenimine; RES, reticuloendothelial system; Tf, transferrin.

by nontarget tissues, inefficient uptake into target cells, and very limited transport into the nucleus of the target cells.

Basically, two classes of nonviral gene transfer systems are being developed: particle-mediated systems, where plasmid DNA is packaged into cationic lipids or polymers; and physical techniques, which are based on application of DNA in combination with a physical device. Administration of naked DNA has already been proven to be an interesting concept for intramuscular vaccination.[2] Physical delivery technologies, such as electroporation[3,4] and the gene gun[5,6] have been found to strongly enhance the efficiency of gene transfer. Although these physical methods are very attractive because of their simplicity of application and the safety aspect of their narrow localization (physical targeting), in many cases the target cells will not be directly accessible. They might be spread throughout the organism, as in the case of cancer metastases, and can be reached only via the systemic route of application. For these instances, particle-based gene transfer systems seem to be more applicable. They can be divided into basically two subclasses: polymer-based transfection complexes named DNA polyplexes, and lipid-based transfection particles named DNA lipoplexes.[7] For systemic in vivo gene delivery the DNA has to be protected from degradation in the biological environment. A sufficiently long circulation time in the blood, and the ability for extravasation and diffusion through the tissues to reach the target site are additional prerequisites. Furthermore, transfection complexes have to be inert against unspecific interactions with biological fluid, extracellular matrix and nontarget cells, but should enable specific binding and uptake by the target cells. Finally, after successful uptake into the target cell, the DNA complexes have to escape from the endosomal compartment to avoid lysosomal degradation, and the DNA has to be imported into the nucleus.

Gene expression after systemic delivery has already been demonstrated with various DNA lipoplexes[8–14] and polyplexes.[15–26] This chapter focuses on the recent findings using DNA polyplexes, with a specific emphasis on tumor-targeted gene transfer.

DNA Polyplexes

A number of positively charged polymers (polycations) have been developed that can interact with the negative phosphate groups of naked DNA, condensing DNA into compact complexes (polyplexes). Polycations include natural DNA binding molecules such as histones,[27] high mobility group proteins,[27] or protamines;[28] the synthetic amino acid homopolymers polylysine,[29] polyarginine,[30] or polyornithine;[31] carbohydrate-based diethylaminoethyl (DEAE)-dextran[32] and chitosan;[33] polymers such as poly(*N*-alkyl-4-vinylpyridinium) salts,[34] poly(2-(dimethylamino)ethyl methacrylate,[35] or polyallylamine derivatives;[36] and protonable polyamines like polyethylenimines,[37] or dendrimers;[38] and histidinylated polylysine;[39] or a series of other modified polymers reviewed in ref.[40] The compaction of DNA by the polycation protects the DNA from nuclease damage and enhances cellular uptake of these compact particles via adsorptive endocytosis or phagocytosis. The extent of DNA compaction with the polycation will depend on the charge ratio,[7] usually defined as the molar ratio of positive charge of polymer to negative charge of the DNA phosphate groups. In the case of polyethylenimines (PEI, where protonation of amine groups depends on the pH), the charge ratio is defined as the molar *N*/*P* ratio of PEI nitrogens to DNA phosphates. When low amounts of polycation are used to complex DNA, around electroneutrality the complexes are usually large and form aggregates. At high *N*/*P* charge ratios polyplexes are soluble, small and positively charged. These polyplexes will interact directly with the cell surface of many cell types due to the presence of negatively charged cell-surface molecules like proteoglycans.[41]

One of the first polycations characterized as a potential nonviral vector for DNA condensation was poly (L)-lysine. Although polylysine can initiate DNA uptake into cells, gene expression is often low due to a number of factors; a major disadvantage with polylysine is that it does not has intrinsic endosomal release activity. Hence, complexes after endocytosis are often not released from endocytic vesicles into the cytoplasm. The enhancement of endosomal release of polylysine/DNA complexes has been achieved via several strategies. The addition of lysosomotropic agents (e.g. chloroquine) into the cell medium can increase gene expression in a number of cell lines.[42] Another approach developed to enhance endosomal release is the incorporation of inactivated viruses[43,44] that have an inherent endosomal disruption mechanism, thus initiating endosomal release of the polylysine complexes. Instead of the whole virus, viral peptide sequences characterized for endosomal activity have been used, such as the N-terminal of rhinovirus VP-1,[44] hemagglutinin 2 subunit,[45] or others.

Some protonable polycations, most prominently the polyamidoamine dendrimers and polyethylenimines (PEI), have been found to be much more effective in transfection than polylysine. There is good evidence that these molecules possess intrinsic endosomal activity.[46] When DNA complexes are taken up by the cell into endosomes and the endosome is acidified, the polycationic polymer acts as a proton sponge. It is thought that polymer protonation as well as influx of counter-ions and water trigger osmotic effects within the endosome, leading to vesicle rupture and release of DNA complexes into the cytoplasm. Because of the high efficiency and low cost of PEI our recent studies have focused on this class of molecules (see below).

Receptor-Binding Ligands Enhance Gene Transfer to Tumor Cells

The design of targeted gene transfer systems is based on DNA/ligand-polycation complexes with cell-binding ligands for specific cell attachment and increased uptake by receptor-mediated endocytosis. This concept has been evaluated by many investigators using various proteins,[15,29,42,47] peptides,[29,47] antibodies,[29,47–49] or carbohydrates[16,29,47] (Fig. 7.1). For targeting of tumor cells we evaluated transferrin (Tf) and epidermal growth factor (EGF) as receptor-binding ligands, covalently conjugated to polyethylenimine (PEI) as a polycation with proven high gene transfer efficiency (see above).

Incorporation of EGF or Tf into DNA/PEI complexes results in a strong (up to 100-fold) enhancement of luciferase gene expression in tumor cell lines, such as erythroleukemic K562 cells or N2A neuroblastoma cells in the case of Tf ligand,[50] or the squamous carcinoma KB cells in the case of EGF ligand.[51] Another consequence of the receptor ligand is the far lower amount of DNA complex required for the transfection. Addition of free, unconjugated ligand to the transfection medium blocks the enhanced expression of ligand-PEI/DNA complexes.[50,51] The size of polyplexes strongly influences the transfection characteristics. In many cases, as reported in,[51,52] the application of rather large particles (0.3–1 μm, generated in Hepes buffered salt, HBS) resulted in higher in vitro transfection activity as compared with small (~50 nm) DNA particles (generated in Hepes buffered glucose, HBG). EGF receptor targeting has also been demonstrated by other investigators using different DNA/polycation complexes.[49,53,54]

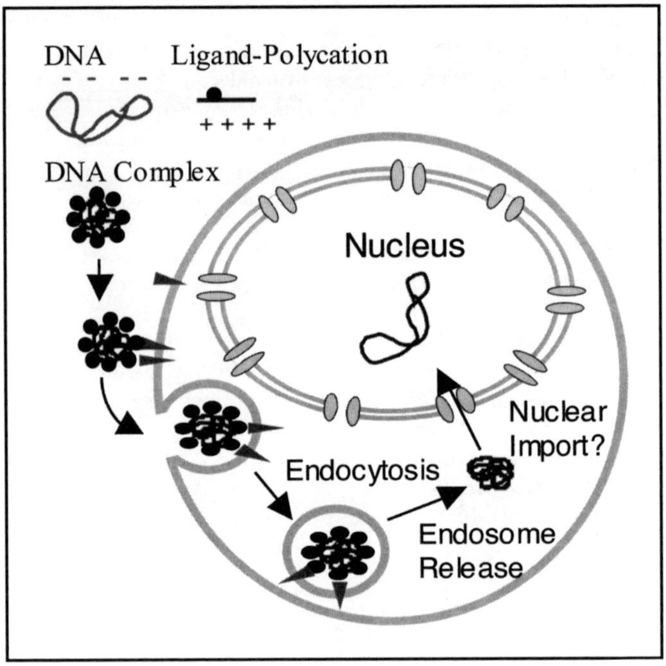

Fig. 7.1. Receptor-mediated delivery of DNA/ligand-polycation complexes.

Nuclear Uptake Barrier: Preferential Gene Transfer to Dividing Cells

Apart from the tumor tissue, the majority of cells of an organism are nondividing. It has been suggested that in lipid-based gene transfer entry of plasmid into the nucleus is the limiting step.[55,56] For polymer-based gene transfer systems (polyfection), the dependence on cell cycle was less clear; there is evidence that nondividing tissues can be transfected in some cases. Therefore the influence of cell cycle on the transfection efficiency by various different methods (lipofection, polyfection, recombinant adenovirus) was investigated.[57] With a centrifugal elutriation technique, K562 cells were fractionated into discrete cell cycle phase specific populations and then transfected (Fig. 7.2). As a positive control, transduction with recombinant adenovirus (which is known to exploit an active nuclear import mechanism) was found to be almost independent of cell cycle (Fig. 7.2A). In contrast, a strong dependence of gene expression on the cell cycle stage was found with Tf-PEI based gene transfer (Fig. 7.2B) and other tested nonviral methods.[57] Transfection efficiency was highest after transfection in late S/G2 phase, shortly before the next mitosis when the nuclear envelope breaks down. These findings indicate that nonviral vectors are limited by the barrier of the nuclear envelope in nondividing cells. However, this mechanism might be exploited in cancer gene therapy concepts to provide an additional level of specificity, because DNA can be effectively incorporated into nuclei of actively dividing cancer cells.

DNA Polyplexes for Systemic Administration

For in vivo gene delivery a series of additional extracellular barriers have to be considered: the interactions of transfection particles with plasma proteins, erythrocytes or

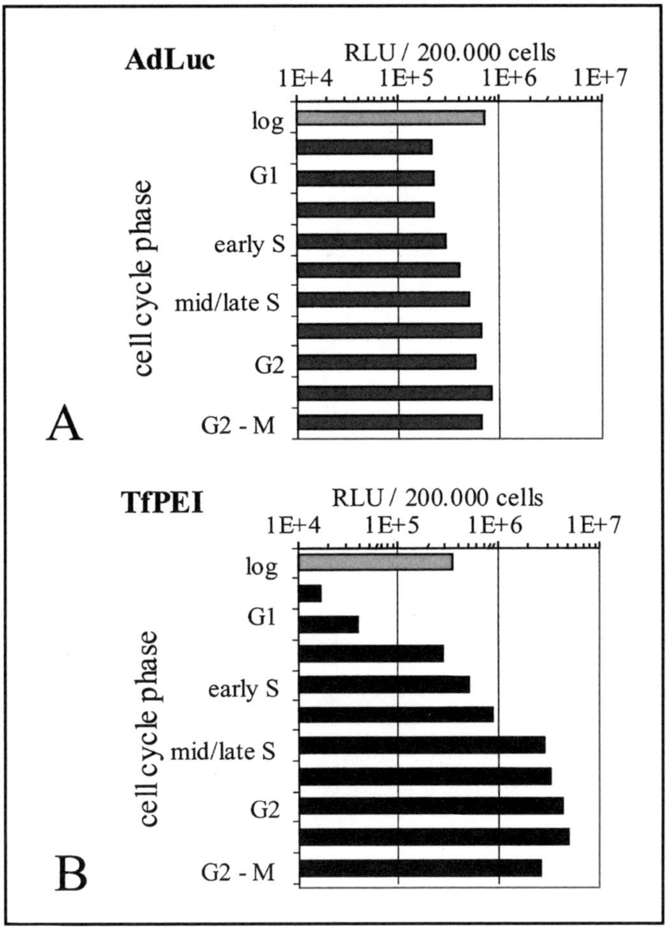

Fig. 7.2. Cell cycle dependence of gene transfer. K562 cells were fractionated by centrifugal elutriation and the individual fractions were transfected with recombinant adenovirus AdLuc (top panel) or Tf-PEI_{800}/pCMVL plasmid complexes (bottom panel). For each time point, 200,000 cells were transfected with 1.3 μg complexed DNA or with 2×10^9 adenoviral particles. Cells were harvested and analyzed for luciferase expression.

nontarget tissue; the aggregation and clearance of transfection particles by the reticuloendothelial system (RES); the restricted extravasation of transfection particles out of the blood stream, and limited diffusion into the target tissue. Despite these limitations, several polyplex systems have already been used for systemic application, demonstrating the first encouraging evidence of in vivo gene expression.

The field has been pioneered by Wu and Wu[15] who applied asialoorosomucoid-polylysine DNA complexes for systemic targeting of hepatocytes which carry asialoglycoprotein receptors. Targeted gene expression was further increased by partial hepatectomy after complexes administration into the rats or via disruption of microtubules with colchicine.[58,59] Genes encoding human albumin and low-density lipoprotein were delivered specifically to livers in rat and rabbit models.[60,61] Related work was published by other investigators[16] using galactose-polylysine for targeting the same liver receptors.

Beside gene transfer to the liver, gene delivery to the lung has been explored using cationic lipoplexes[8–11] and, more recently, polyplexes. Systemically delivered PEI/DNA complexes resulted in a high gene expression in the lung with lower activity gene expression being observed in other organs.[17–19] These complexes rapidly crossed the endothelium of the lung capillary bed, and gene expression was mainly found in alveolar cells. The high lung transfection activity was observed only with linear PEI22 of low molecular weight of 22 kDa. Complexes had to be generated under salt-free conditions (to form small particles) and, to provide optimum results, had to be applied in a rather large volume (0.4 mL) of osmotically balanced glucose solution to avoid mixing with blood.[19] Branched PEIs (25 kDa and 800 kDa) were less efficient and more toxic as compared to the linear PEI.[17] Though the positive charge of polyplexes seems to be a useful feature for interacting with the lung tissue, it is also one obvious reason for unspecific interactions with plasma components or erythrocytes. Positive complexes containing polylysine, PEI or dendrimer are very active stimulators of the plasma complement system.[62] Other plasma components, such as IgM, fibrinogen, fibronectin or albumin, also were shown either to coat the DNA complexes[20,23] causing either aggregation of the complexes, or to initiate dissociation resulting in degradation of the DNA. Interactions with complement[62] and with cells of the RES[63] lead to rapid destruction and removal of DNA complexes, and can cause a variety of undesired effects associated with high level of complement or RES activation. However, recently several methods have been successfully applied to shield the positive surface charge of polyplexes, reducing nonspecific interactions and aggregation behavior of the complexes.

We have shielded the surface of DNA/Tf-PEI_{800} complexes by covalent modification with polyethylene glycol (PEG) as illustrated in Fig. 7.3 (*center*). Without shielding, the high molecular weight PEI (800 kDa) used in this study displays high in vivo toxicity. The PEG shielding can prevent the unspecific interaction with plasma compounds or erythrocytes, and strongly reduces the toxicity of the complexes after systemic application.[20] Shielded PEI/DNA complexes have increased circulation time before being cleared from the blood stream as compared to unshielded complexes. The shield does not interfere with the transferrin-mediated specific interactions.[20] A similar observation was made when EGF was employed as ligand.[51] Similarly, PEG-polylysine/DNA shielded complexes had decreased surface charge and reduced toxicity. Solubility was increased as compared to unshielded polylysine/DNA complexes.[64] Copolymers of anionic peptides have been linked to PEG which stabilize cationic complexes by inhibiting aggregation and reducing complement activation.[25]

Other compounds used for shielding of polyplexes are Poly-*N*-(2-hydroxypropyl) methacrylamide (pHPMA) and poloxamer. Polylysine grafted with pHPMA had increased solubility with lower toxicity and a negative surface charge. Furthermore, the pHPMA shield reduced the interaction of complexes with blood components.[24] PEI/DNA complexes shielded with the polyether Pluronic 123 showed reduced gene expression in the lung with an increased expression in the liver over nonshielded PEI complexes.[26] Local application of DNA shielded with poloxamers increased gene expression in muscle.[65]

Another strategy for masking the surface charge of DNA complexes which we employed recently is shown in Fig. 7.3 (*right*). Transferrin is not only a very effective receptor ligand but also a plasma protein which by nature might present an effective shielding agent. In fact, efficient charge shielding was obtained by incorporation of sufficient Tf ligand to DNA complexes with the lower molecular weight (25 kDa) PEI_{25}, even in the absence of PEGylation.[22] A similar concept has already been applied for shielding cationic lipoplexes with noncovalently associated transferrin.[13] With DNA/PEI complexes, transferrin shielding was only effective with covalent linkage to the polymer.[22]

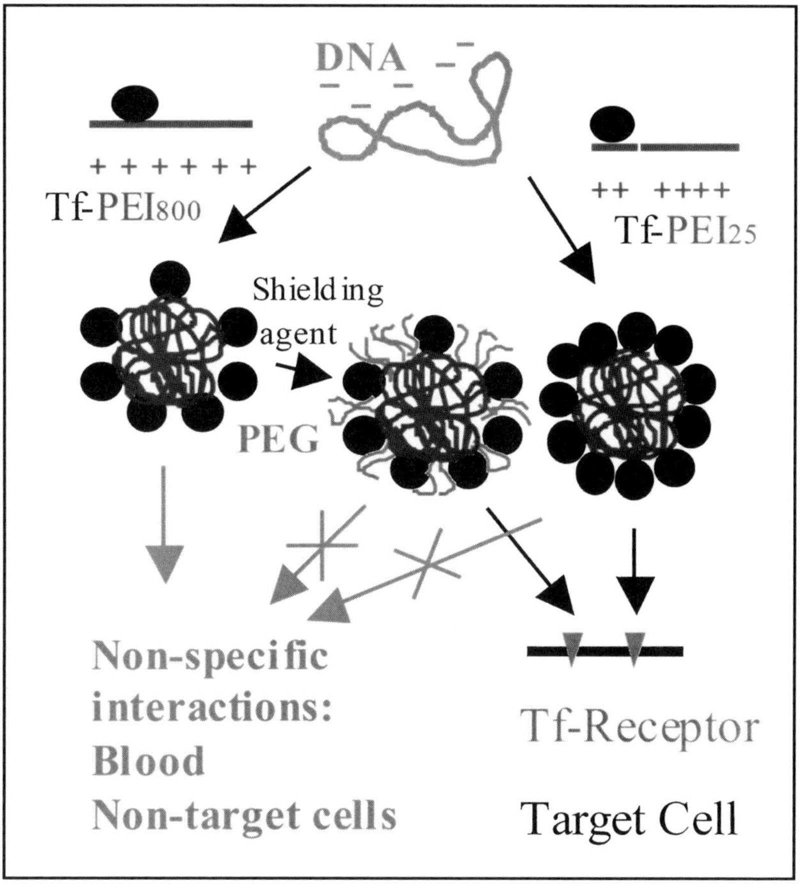

Fig. 7.3. Surface-shielded transferrin-PEI/DNA complexes for systemic gene delivery. Positively charged DNA/polycation complexes can interact by nonspecific electrostatic interaction with blood components and with any negatively charged cell surface (bottom, left). These nonspecific interactions can be blocked by shielding the surface by PEGylation, still allowing specific binding to the target provided that the ligand is not hidden by the PEG coat (bottom, center). Similarly, complexes which have the ligand transferrin at sufficiently high densities are shielded from nonspecific interactions but allowing specific binding to the transferrin receptor (bottom, right).

A novel mechanism for stealthing gene delivery particles has recently been proposed by the group of Jon Wolff[66] who studied the interactions of peptides with the innate immune system using T7 Phage display. This work identified peptide sequences which bind to serum proteins such as CRP or alpha-macroglobulin. Coating with these proteins helps to evade the innate immune system. Incorporation of such protective peptide sequences into polyplexes remains to be shown as useful.

In Vivo Gene Transfer in Tumor-Bearing Mice

DNA polyplexes have already been used for local administration to tumors.[67–69] For evaluating systemic gene delivery with shielded PEI polyplexes, we chose a subcutaneous tumor model, using a well-vascularized, moderately rapid growing tumor (Neuro2A

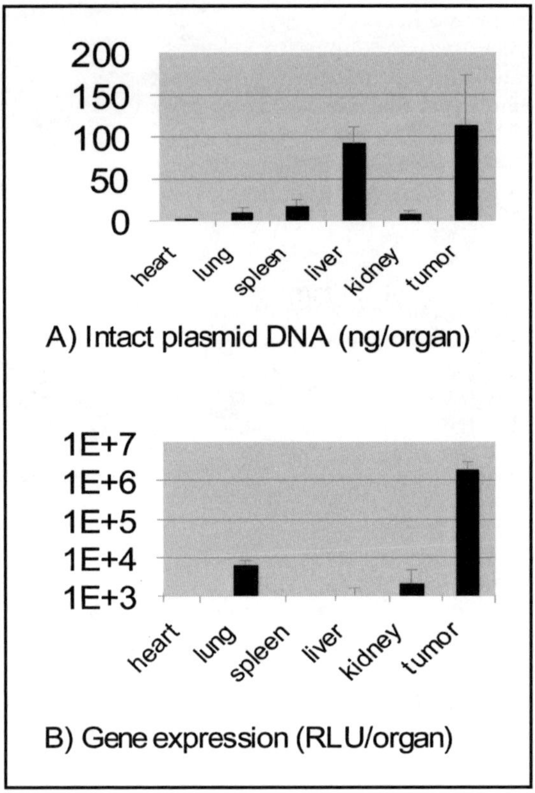

Fig. 7.4. Targeting of gene expression to distant tumors after systemic application. Surface-shielded transferrin-PEI/DNA complexes (containing 50 μg pCMVL plasmid) are injected via the tail vein into the systemic blood circulation of mice bearing subcutaneous neuroblastoma tumors. This results in delivery of DNA particles into the distant tumor (top) and targeted gene expression in the tumor as shown for luciferase gene expression (bottom).

neuroblastoma in A/J mice). Different formulations (containing 50 μg of the pCMVL plasmid carrying the luciferase reporter gene) were injected into the tail vein, and reporter gene expression was detected within the different tissues.[20,21] After injection of naked DNA, luciferase expression was found at the injection site only, but not in distant organs.[21] Positively charged PEI_{800}/DNA and TF-PEI_{800}/DNA complexes resulted in expression primarily in the lung, but not in the tumor.[20,21] In sharp contrast, electroneutral formulations of PEG/Tf-PEI_{800}/DNA (shielded by PEG) or Tf-PEI_{25}+PEI_{25}/DNA (shielded by transferrin, see Fig. 7.4) resulted in preferential gene transfer to the distant tumor.[20–22] The tumor-selective expression levels (up to 100-fold higher than in other organs including the lungs) were produced by complexes that were not immediately cleared from the circulation. Considerable amounts of intact plasmid DNA were found in the distant tumor (Fig. 7.4A). The selectivity of expression (Fig. 7.4B) could also be the consequence of the specific physiology of tumor tissue: higher permeability of the tumor vasculature, and the presence of dividing (transfection-competent) tumor cells.

Within tumors, plasmid DNA and gene expression were found at several sites throughout the tumor, but not homogenously distributed within the tumor tissue. It appeared that the spots of strong expression often were close to immature blood vessels,[22] where DNA seemed to be able to extravasate only a few cell layers. Consequently, with the current type of polyplex vectors, therapeutic effects might only be expected with genes expressing very potent biomolecules that can act also in paracrine fashion (displaying the effect on surrounding cells). Therefore, we tested targeting of shielded transferrin-PEI/DNA complexes containing the expression plasmid for tumor necrosis factor α (TNFα) to distant tumors following systemic application. Initial results were very encouraging, showing hemorrhagic tumor necrosis and reduction of tumor growth without TNFα-related systemic side effects.[70–71] Further investigations will evaluate the therapeutic usefulness of the concept.

In addition to demonstrating tumor-directed gene delivery using polyplexes, investigators have generated tumor-targeted lipoplexes[13,14] in a similar fashion. In these developments, the shielding of DNA complexes by molecules such as PEG[14] or transferrin[13] has been found to be crucial for tumor targeting.

Conclusions

Both, biological parameters (such as appropriate targeting ligand, relevant therapeutic gene, and expression cassette) and biophysical parameters of transfection complexes (such as size, surface charge, molecular weight of the polymeric carrier, modification with targeting ligand, shielding agent) can determine distribution of gene expression and therapeutic outcome. Using electroneutral DNA complexes, targeted gene delivery to a distant tumor is possible. We consider the transfection systems that we applied as a first "proof-of-concept" step; further improvements are necessary for clinical applications such as cancer gene therapy. These improvements include the generation of DNA formulations of sufficiently small sizes to distribute within the target organ, formulations that are stable for storage but still very potent when reaching the target cell, and the incorporation of target-specific transcription elements.

Acknowledgment

The contributions of Thomas Blessing, Sylvia Brunner, Robert Holzhauser, Ralf Kircheis, Malgorzata Kursa, Vanessa Rössler, Susanne Schüller, Lionel Wightman, and Wolfgang Zauner are greatly appreciated. Some of the reported work was supported by the Austrian Science Foundation grant S07405-MOB, the EC grant BIO4 CT97-2191, and Boehringer Ingelheim, Austria.

Literature Cited

1. Lollo CP, Banaszczyk MG, Chiou HC. Curr Opin Mol Ther 2000; 2:136-142.
2. Wang R. Science 1998; 282:476-480.
3. Mathiesen I. Gene Ther 1999; 6:508-514.
4. Widera G. J Immunol 2000; 164:4635-4640.
5. Yang N-S, Burkholder J, Roberts B, Martinell B, McCabe D. Proc Natl Acad Sci USA 1990; 87:9568-9572.
6. Kuriyama S, Mitoro A, Tsujinoue H, Nakatani T, Yoshiji H, Tsujimoto T, Yamazaki M, Fukui H. Gene Ther 2000; 7:1132-1136.
7. Felgner PL. Hum Gene Ther 1997; 8:511-512.
8. Hong K, Zheng W, Baker A, Papahadjopoulos D. FEBS Letters 1996; 400:233-237.
9. Li S, Huang L. Gene Ther 1997; 4:891-900.
10. Liu F, Qi H, Huang L, Liu D. Gene Ther 1997; 4:517-523.
11. Liu Y. Nat Biotechnol 1997; 15:167-173.

12. Templeton NS, Lasic DD, Frederik PM, Strey HH, Roberts DD, Pavlakis GN. Nat Biotechnol 1997; 15:647-652.
13. Xu L, Pirollo KF, Chang EH. Hum Gene Ther 1999; 10:2941-2952.
14. Monck MA. J Drug Target 2000; 7:439-452.
15. Wu GY, Wu CH. J Biol Chem 1988; 263:14621-14624.
16. Perales JC, Ferkol T, Beegen, H, Ratnoff OD, Hanson RW. Proc Natl Acad Sci USA 1994; 91:4086-4090.
17. Goula D, Benoist D, Mantero S, Merlo G, Levi G, Demeneix BA. Gene Ther 1998; 5:1291-1295.
18. Goula D, Becker N, Lemkine GF, Normandie P, Rodrigues J, Mantero S, Levi G, Demeneix BA. Gene Ther 2000; 7:499-504.
19. Zou SM, Erbacher P, Remy JS, Behr JP. J Gene Med 2000; 2:128-134.
20. Ogris M, Brunner S, Schüller S, Kircheis R, Wagner E. Gene Ther 1999; 6:595-605.
21. Kircheis R, Schueller S, Brunner S, Ogris M, Heider K-H, Zauner W, Wagner E. J Gene Med 1999; 1:111-120.
22. Kircheis R, Wightman L, Schreiber A, Robitza B, Rössler V, Kursa M, Wagner E. Gene Ther 2001; 8:28-40.
23. Dash PR, Read ML, Barrett LB, Wolfert MA, Seymour LW. Gene Ther 1999; 6:643-650.
24. Dash PR, Read ML, Fisher KD, Howard KA, Wolfert M, Oupicky D, Subr V, Strohalm J, Ulbrich K, Seymour LW. J Biol Chem 2000; 275:3793-3802.
25. Finsinger D, Remy JS, Erbacher P, Koch C, Plank C. Gene Ther 2000; 7:1183-1192.
26. Nguyen HK, Lemieux P, Vinogradov SV, Gebhart CL, Guérin N, Paradis G, Bronich TK, Alakhov VY, Kabanov AV. Gene Ther 2000; 7:126-138.
27. Zaitsev SV, Haberland A, Otto A, Vorobev VI, Haller H, Böttger M. Gene Ther 1997; 4:586-592.
28. Wagner E, Cotten M, Foisner R, Birnstiel ML. Proc Natl Acad Sci USA 1991; 88:4255-4259.
29. Zauner W, Ogris M, Wagner E. Adv Drug Del Rev 1998; 30:97-113.
30. Emi N, Kidoaki S, Yoshikawa K, Saito H. Biochem Biophys Res Commun 1997; 231:421-424.
31. Dong Y, Skoultchi AI, Pollard JW. Nucleic Acids Res 1993; 21:771-772.
32. Lopata MA, Cleveland DW, Sollner WB. Nucleic Acids Res 1984; 12:5707-5717.
33. Erbacher P, Zou S, Bettinger T, Steffan AM, Remy JS. Pharm Res 1998; 15:1332-1339.
34. Kabanov AV, Astafieva IV, Maksimova IV, Lukanidin EM, Georgiev GP, Kabanov VA. Bioconjug Chem 1993; 4:448-454.
35. van de Weteringen P, Cherng JY, Talsma H, Crommelin DJ, Hennink WE. J Control Rel 1998; 53:145-153.
36. Boussif O, Delair T, Brua C, Veron L, Pavirani A, Kolbe HV. Bioconjug Chem 1999; 10:877-883.
37. Boussif O, Lezoualc'h F, Zanta MA, Mergny MD, Scherman D, Demeneix B, Behr JP. Proc Natl Acad Sci USA 1995; 92:7297-7301.
38. Haensler J, Szoka FC. Bioconjug Chem 1993; 4:372-379.
39. Midoux P, Monsigny M. Bioconjug Chem 1999; 10:406-411.
40. Han SO, Mahato RI, Sung YK. Mol Ther 2000; 2:302-317.
41. Mislick KA, Baldeschwieler JD. Proc Natl Acad Sci USA 1996; 93:12349-12354.
42. Wagner E, Curiel D, Cotten M. Adv Drug Del Rev 1994; 14:113-135.
43. Wagner E, Zatloukal K, Cotten M, Kirlappos H, Mechtler K, Curiel DT, Birnstiel ML. Proc Natl Acad Sci USA 1992; 89:6099-6103.
44. Zauner W, Blaas D, Küchler E, Wagner E. J Virol 1995; 69:1085-1092.
45. Wagner E, Plank C, Zatloukal K, Cotten M, Birnstiel ML. Proc Natl Acad Sci USA 1992; 89:7934-7938.
46. Kichler A, Behr J-P, Erbacher P. In Nonviral Vectors for Gene Therapy, L Huang, MC Hung, E Wagner (eds.), Academic Press, 1999; New York.
47. Wagner E. In Nonviral Vectors for Gene Therapy, L Huang, MC Hung, E Wagner (eds.), Academic Press, New York, 1999, pp. 208-227.
48. Buschle M. Hum Gene Ther 1995; 6:753-761.
49. Chen J, Gamou S, Takayanagi A, Shimizu N. FEBS Lett 1994; 338:167-169.
50. Kircheis R, Kichler A, Wallner G, Kursa M, Ogris M, Felzmnn T, Buchberger M, Wagner E. Gene Ther 1997; 4:409-418.
51. Blessing, T, Kursa M, Holzhauser R, Kircheis R, Wagner E. Different strategies for Formation of PEGylated EGF-conjugated PEI/DNA complexes for targeted gene delivery. Bioconjug Chem 2001; 12:529-537.
52. Ogris M, Steinlein P, Kursa M, Mechtler K, Kircheis R, Wagner E. Gene Ther 1998; 5:1425-1433.
53. Shimizu N, Chen J, Gamou S, Takayanagi A. Cancer Gene Ther 1996; 3:113-120.
54. Cristiano RJ, Roth JA. Cancer Gene Ther 1996; 3:4-10.

55. Zabner J, Fasbender AJ, Moninger T, Poellinger KA, Welsh MJ. J Biol Chem 1995; 270:18997-19007.
56. Brisson M, Huang L. Curr Opin Mol Ther 1999; 1:140-146.
57. Brunner S, Sauer T, Carotta S, Cotten M, Saltik M, Wagner E. Gene Ther 2000; 7:401-407.
58. Chowdhury NR, Wu CH, Wu GY, Yerneni PC, Bommineni VR, Chowdhury JR. J Biol Chem 1993; 268:11265-11271.
59. Chowdhury NR, Hays RM, Bommineni VR, Franki N, Chowdhry JR. J Biol Chem 1996; 271:2341-2346.
60. Wu GY, Wilson JM, Shalaby F, Grossman M, Shafritz DA, Wu CH. J Biol Chem 1991; 266:14338-14342.
61. Wilson JM, Grossman M, Wu CH, Chowdhury NR, Wu GY, Chowdhury JR. J Biol Chem 1992; 267:963-967.
62. Plank C, Mechtler K, Szoka F, Wagner E. Hum Gene Ther 1996; 7:1437-1446.
63. Mahato RI, Kawabata K, Nomura T, Takakura Y, Hashida M. J Pharmacol Sci 1995; 84:1267-1271.
64. Toncheva V, Wolfert MA, Dash PR, Oupicky D, Ulbrick K, Seymour LW, Schacht EH. Biochim Biophys Acta 1998; 1380:354-368.
65. Lemieux P, Guerin N, Paradis G, Proulx R, Chistyakova L, Kabanov A, Alakhov V. Gene Ther 2000; 7:986-991.
66. Sokoloff AV, Bock I, Zhang G, Sebestyen MG, Wolff JA. Mol Ther 2000; 2:131-139.
67. Coll JL, Chollet P, Brambilla E, Desplanques D, Behr JP, Favrot M. Hum Gene Ther 1999; 10:1659-1666.
68. Maheshwari A, Mahato RI, McGregor J, Han S, Samlowski WE, Park JS, Kim SW. Mol Ther 2000; 2:121-130.
69. Gautam A, Waldrep JC, Densmore CL, Koshkina N, Melton S, Roberts L, Gilbert B, Knight V. Gen Ther 2002; 9:353-357.

Chapter 8

Photochemical Transfection: A Technology for Efficient Light-Directed Gene Delivery

Anders Høgset, Lina Prasmickaite, Marit Hellum, Birgit Ø. Engesæter, Vibeke M. Olsen, Torunn E. Tjelle, Carl J. Wheeler, and Kristian Berg

Abstract

Most synthetic gene delivery vectors are taken up in the cell by endocytosis, and inefficient escape of the transgene from endocytic vesicles often is a major barrier for gene transfer by such vectors. To improve endosomal release we have developed a new technology, named photochemical internalization (PCI). PCI is based on photochemical reactions initiated by photosensitizing compounds localized in endocytic vesicles, inducing rupture of these vesicles upon light exposure. PCI constitutes an efficient light-inducible gene transfer method *in vitro*, which potentially can be developed into a site-specific method for gene delivery in *in vivo* gene therapy. In this paper the principle behind the PCI technology and the effect of PCI on transfection with different synthetic gene delivery vectors are reviewed. PCI treatment by the photosensitizer aluminum phthalocyanine ($AlPcS_{2a}$) strongly improves transfection mediated by cationic polymers (e.g., poly-L-lysine and polyethylenimine), while the effect on transfection with cationic lipids is more variable. The timing of the light treatment relative to the transfection period was also important, indicating that release of the DNA from early endosomes is important for the outcome of PCI-induced transfection. The possibilities of using PCI as a technology for efficient, site-specific gene delivery in *in vivo* gene therapy is discussed.

Introduction

Background

The development of technologies for efficient and specific delivery of therapeutic genes to the target cells remains a main issue in the clinical development of gene therapy.[1,2] Especially, despite continuous improvement in vector technology, delivery of functional genes by synthetic vectors is still a very inefficient process. While for some viral vectors the cellular uptake of just a few virus particles per cell may be sufficient to obtain

Synthetic DNA Delivery Systems, edited by Dan Luo and W. Mark Saltzman.

transduction,[3,4] with synthetic vectors cells may not be transfected even though they contain large amounts of exogenous DNA. For example in different systems it has been estimated that the cells on average contain 2×10^5 to 3×10^5 plasmids per cell after delivery by cationic lipids,[5,6] or about 6×10^4 plasmids per cell after polylysine-mediated delivery.[7] Therefore, for synthetic vectors the process of intracellular transport of the transgene must provide severe barriers for functional delivery of transgenes. Nuclear transport and uptake of the transfecting DNA represent such important barriers.[5,8–10] However, several other intracellular barriers can also be identified, e.g., dissociation of DNA from the complexing agent, degradation of the DNA, and sequestration of the DNA in endocytic vesicles.[5,11–13] With most synthetic gene delivery vectors, the major mechanism of uptake of gene complexes is through the endocytic pathway,[5,10,14–16] and in order for the transfecting DNA to be translocated to the nucleus it has to be released from the endocytic vesicles before being degraded by lysosomal enzymes. The importance of endosomal escape for the outcome of transfection has been indicated by several lines of evidence: (i) lysosomotropic agents such as chloroquine can substantially improve transfection by several synthetic transfection agents,[17–20] although the mechanism behind this effect is not fully clear;[21] (ii) the inclusion of inactivated adenovirus with, e.g., polylysine–DNA complexes enhances transfection considerably, most probably due to endosomolytic properties of the virus proteins;[18,22–24] (iii) several "fusogenic" peptides have the same effect;[11,19] (iv) glycerol can enhance transfection, presumably through an increase in endosomal permeability;[25,26] (v) the inclusion of "fusogenic" lipids in lipoplexes often stimulates transfection.[27–29]

While some polycations, such as polylysine, probably have no inherent mechanism for endosomal escape, it has been proposed that, e.g., polyethylenimine can effect endosomal escape through a "proton sponge" effect leading to swelling and permeabilization of the endosomes.[13,30] Likewise, for cationic lipids, a model has been suggested where the cationic lipids induce a "flip-flop" of phospholipids in the endosomal membranes leading to release of the transfecting DNA both from the lipid complex and from the endosomes.[31,32] These mechanisms are, however, often probably far less efficient than, e.g., viral mechanisms for endosomal escape, implying that more efficient methods for endosomal release could still lead to substantial improvement of transfection with synthetic transfection agents.

We have recently described a technology, named photochemical internalization (PCI), where photosensitizing compounds and light are employed to rupture membranes of endocytic vesicles, resulting in the release of functional macromolecules from these vesicles into the cell cytosol.[33] In this paper the background of this technology, its employment as a gene transfection method, and its possible use as a system for site-specific gene delivery in vivo is reviewed.

Photosensitizers and Photochemical Reactions

General Principles

Photosensitizers (S) are compounds that upon absorption of light (hν) become excited to an excited state (S*), which can initiate further photochemical reactions. With porphyrins and structurally similar compounds the most important of these reactions proceeds via singlet oxygen ($^1O_2$), a highly reactive form of oxygen formed after interaction between a light-activated photosensitizer and triplet ground-state molecular oxygen ($^3O_2$).[34,35] Schematically:

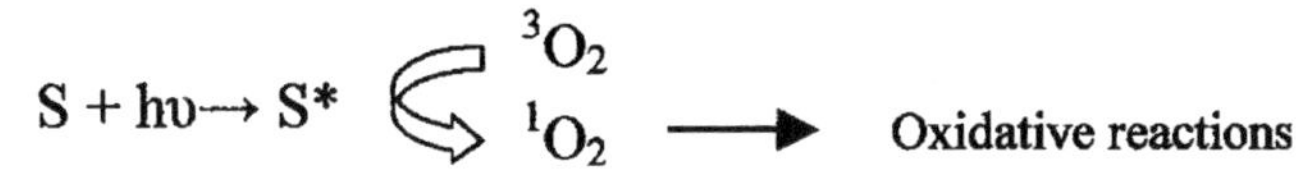

Singlet oxygen is a powerful oxidizing agent that can oxidize various biomolecules such as unsaturated fatty acids, certain amino acids (His, Trp, Cys, Met, and Tyr), and the nucleic acid base guanine.[36] Thus, $^1O_2$ can induce damage into various cellular structures. However, since $^1O_2$ has a very short lifetime (<0.1 μs) and accordingly a short range of action (10–20 nm),[37] only targets very close to the excited photosensitizer will be subject to $^1O_2$ damage upon light exposure, while distant molecules will be left intact.

Photochemically Induced Cytotoxicity and Photodynamic Therapy of Cancer

In general, photochemical treatment induces cytotoxic effects, and this has been successfully exploited for cancer treatment in photodynamic therapy (PDT), a treatment modality where light exposure leads to photosensitizer-induced killing of cancer cells.[38–40] PDT has been approved for the treatment of several cancers, such as basal cell carcinoma and cancers in the lung, esophagus, and bladder, as well as age-related macula degeneration and solar keratosis.

In vivo some photosensitizers show a preferential accumulation in tumor tissue as compared to adjacent normal tissue.[38] This accumulation seems to be related to certain properties of tumor tissue, such as leaky vasculature, low pH, poor lymphatic drainage, or the presence of tumor-infiltrating macrophages taking up aggregated photosensitizers.[38,39]

Intracellular Localization of the Photosensitizers

Depending on their physicochemical properties, different photosensitizers enter the cell in different ways and localize at different sites inside the cell.[41–44] Photosensitizers that aggregate and hydrophilic photosensitizers such as sulfonated tetraphenyl porphines ($TPPS_n$)[43,45] and sulfonated aluminum phthalocyanines ($AlPcS_n$)[41,46,47] enter the cells via endocytosis and tend to accumulate in endocytic vesicles. $TPPS_{2a}$ (*meso*-tetraphenylporphine with two sulfonate groups on adjacent phenyl rings) and $AlPcS_{2a}$ (aluminum phthalocyanine with two sulfonate groups on adjacent phthalate rings) (Fig. 8.1) have two sulfonate groups on adjacent phenyl/phthalate rings, making the photosensitizer molecules amphiphilic. Therefore, they primarily localize in the membranes of endocytic vesicles[45] with their hydrophobic part inserted into the membrane.[48]

The Principle of Photochemical Internalization

We have shown that light exposure of cells containing photosensitizers in their endocytic vesicles leads to a permeabilization of the vesicles and release of the photosensitizer.[45,49] In some cases, e.g., for the photosensitizers $TPPS_{2a}$ and $TPPS_1$ (*meso*-tetraphenylporphine with one sulfonate group), substantial amounts of lysosomal enzyme activities also have been found in the cytosol after light treatment, indicating that molecules located inside the lysosomes can be released into the cytosol without losing their biological

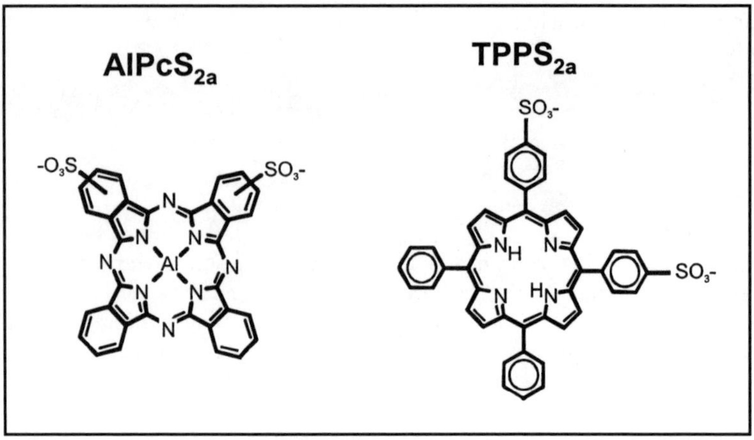

Fig. 8.1. Structures of $AlPcS_{2a}$ and $TPPS_{2a}$.

activity.[45] It has also been found that a large fraction of lysosomes may be ruptured without inducing significant cell death. This has been attributed to a high sensitivity of cathepsins to inactivation by the photochemical treatment.[45] Cathepsins are lysosomal proteases of crucial importance for the cytotoxic effect of lysosomal rupture,[50,51] and their inactivation would be expected to lower the cytotoxic effect of lysosome rupture.

However, even though cathepsins seem to be destroyed, many other vesicular molecules maintain their biological activity also after the photochemical treatment. This is probably due to the short range of action of the photochemically generated $^1O_2$, meaning that only molecules very close to the excited photosensitizer will be damaged. Thus, if the photosensitizer mainly localizes in the vesicle membranes, molecules in the vesicle matrix should not be very susceptible to photochemical damage. Taken together this means that the photochemical treatment may be used to release a host of endocytosed molecules in a biologically active form from endocytic vesicles, a principle we have named photochemical internalization (PCI) (Fig. 8.2A). The introduction of molecules into the cytosol is achieved by first exposing the cells or tissues to a photosensitizing dye and the molecule that one wants to deliver, both of which should preferentially localize in endosomes and/or lysosomes.[33] Secondly, the cells or tissues are exposed to light of wavelengths inducing a photochemical reaction. This reaction will lead to disruption of lysosomal and/or endosomal membranes and the contents of these vesicles will be released into the cytosol. Detailed descriptions of practical aspects of the technology can be found in Berg et al.[53] and Prasmickaite et al.[54]

In vitro, PCI has been shown to induce endosomal release of a variety of molecules, such as plant toxins, an immunotoxin, a peptide, ribozymes, and oligodeoxynucleotides (Fig. 8.2B).[33,52,55,56] In addition, as will be discussed below, it has been shown to substantially increase transfection with several polycationic transfection agents.[7,33,44,52]

Recent experiments on PCI with a protein toxin show that photochemical internalization works very well also in vivo. Thus, the cytotoxic effect of gelonin, which is relatively nontoxic to tumor cells when used alone, was increased tremendously, and in a site-specific way, by the photochemical treatment, resulting in a complete regression of subcutaneous tumors in a mouse model.[57] This result indicates that the PCI technology has a clear clinical potential, both for the light-directed delivery of drugs and, as will be discussed below, for the light-induced, site-specific delivery of genes in in vivo gene therapy.

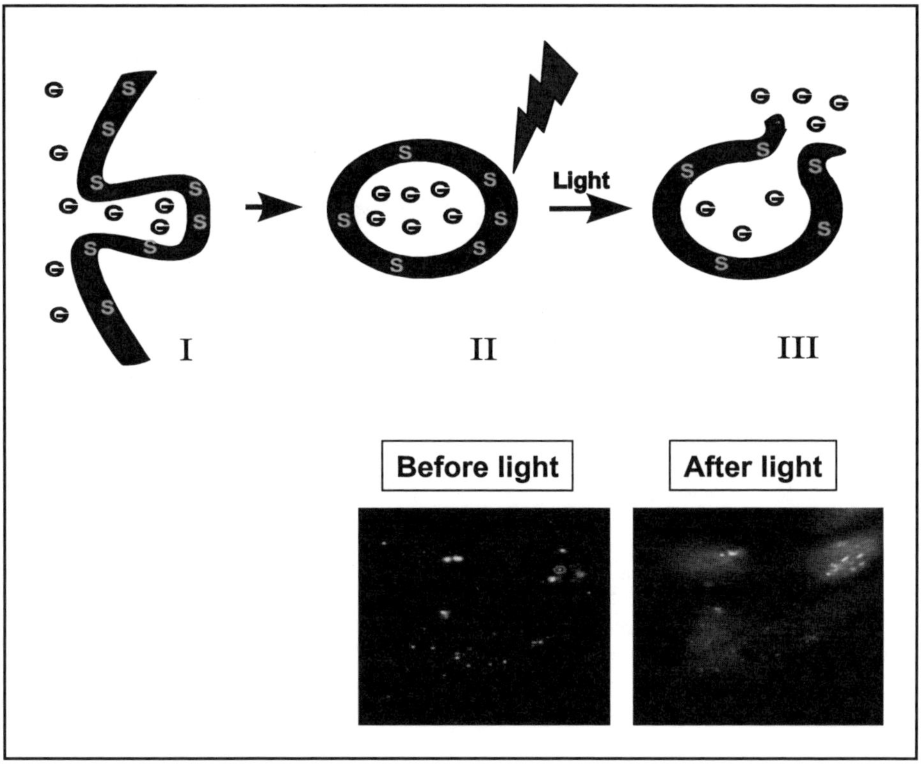

Fig. 8.2. Light-induced cytosolic delivery by PCI. (A) *The principle for PCI.* I. The cells are incubated with the photosensitizer (S) and the molecule to be delivered (G). II. S and G are taken up in the cell by endocytosis, localizing in the endosome membrane and matrix, respectively. III. Illumination induces photochemical damage to the endosome membrane releasing G into the cytosol. (B). *PCI-induced relocalization of an oligodeoxynucleotide.* THX cells were incubated for 18 h with 20 μg/mL $AlPcS_{2a}$. After 3-h wash in culture medium and 4-h incubation with a polylysine complex of a fluorescein-labeled oligodeoxynucleotide the same cells were photographed before and after the exposure to 10 s microscopy light. See Ref. (52) for details. It can be clearly seen that after light treatment the fluorescence spreads out in the entire cell from the spots where it was initially located.

Photochemical Internalization as a Gene Delivery Technology

We have studied photochemical internalization as a gene delivery technology mainly by using an EGFP (enhanced green fluorescent protein)-encoding plasmid and polycationic transfection agents.[33,52] However, as will be discussed below, we have also tested the effect with some cationic lipid vectors.[7,44]

The spectral characteristics of EGFP and the photosensitizer $AlPcS_{2a}$ allow their simultaneous specific detection by fluorescence microscopy. Thus, in initial experiments we used this method to study the effect of PCI on polylysine-mediated transfection of $AlPcS_{2a}$-treated HCT 116 human colon carcinoma cells (Fig. 8.3). A part of the culture dish was covered by aluminum foil, while the rest of the dish was illuminated. As can be seen from Fig. 8.3A a substantial fraction of the irradiated HCT 116 cells exhibited visible EGFP-fluorescence, while only very few of the nonirradiated cells expressed this reporter

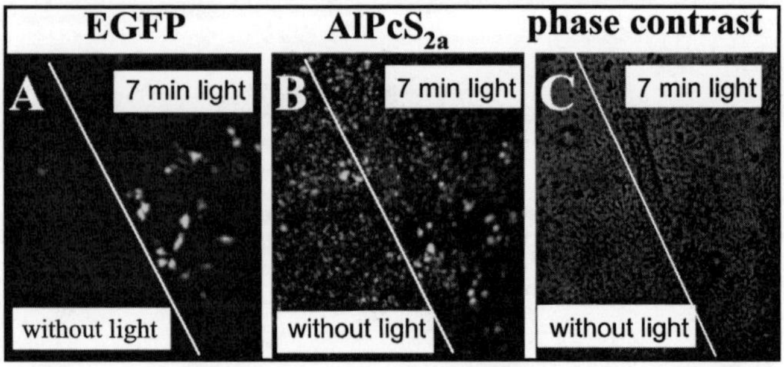

Fig. 8.3. PCI-induced transfection of HCT 116 cells studied by fluorescence microscopy. The cells were incubated with for 18 h with 20 μg/mL $AlPcS_{2a}$, washed three times with culture medium and treated with a pEGFP-N1/polylysine complex for 6 h. The cells were washed once with culture medium, and the culture dish was partly covered by aluminum foil before being subjected to 7-min light treatment. Forty-eight hours later the cells were analyzed by fluorescence microscopy for EGFP (A) and $AlPcS_{2a}$ (B) fluorescence, as well as by phase contrast microscopy (C), as indicated on the figure (see Ref. (52) for details).

protein. This shows that the light treatment can induce transfection in HCT 116 cells, and also indicates that a highly site-specific induction can be obtained, reflected by the clear difference in transfection between the illuminated and the nonilluminated parts of the culture dish. The $AlPcS_{2a}$ fluorescence (Fig. 8.3B) shows that the number of cells containing $AlPcS_{2a}$ was roughly the same in the illuminated and the nonilluminated parts of the dish.

Light Dose Dependence

For more exact quantification of the effects of PCI on transfection, flow cytometry was used for analysis of the transfected cells. From Fig. 8.4 it can clearly be seen that the light treatment induces a substantial increase in the transfection efficiency for THX melanoma

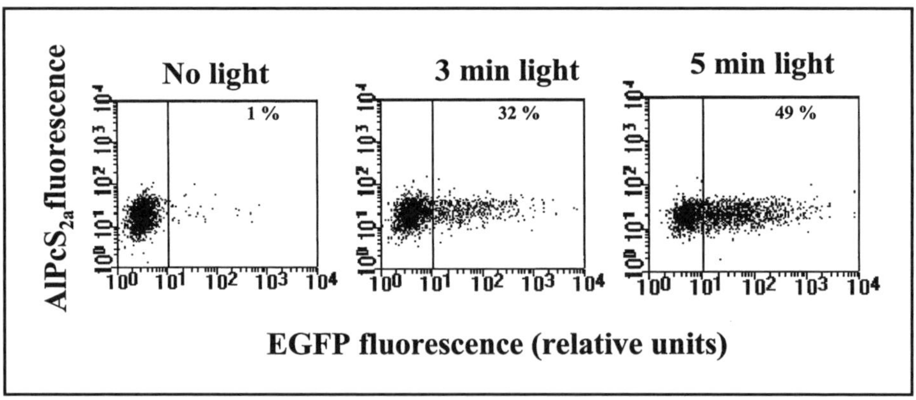

Fig. 8.4. Light-induced transfection studied by flow cytometry. THX-cells were treated with 20 μg/mL $AlPcS_{2a}$ for 18 h, washed in photosensitizer-free medium, and incubated with a pEGFP-N1/polylysine complex (5 μg/mL DNA) essentially as described in the legend to Fig. 8.3. The cells were exposed to light (illumination times indicated on the figure), and red ($AlPcS_{2a}$) and green (EGFP) fluorescence were analyzed by flow cytometry as described earlier.[52] The cells to the right of the vertical line were taken as positive for EGFP-expression, since in control experiments (cells that received no plasmid) there were no cells in this area (not shown).

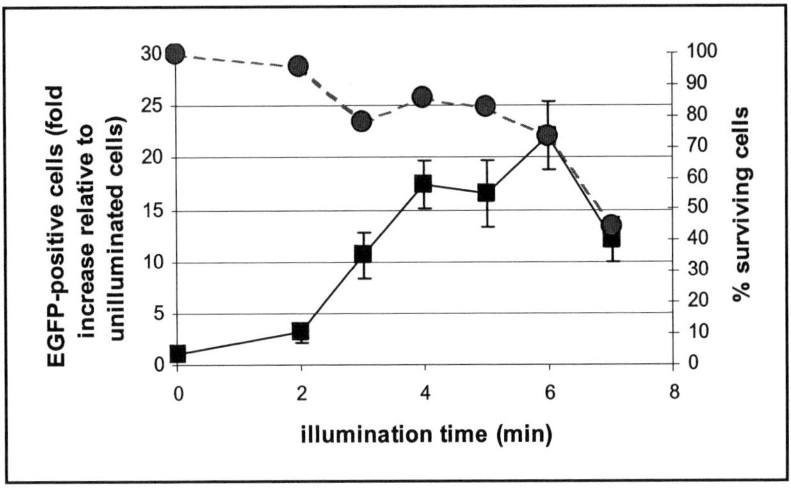

Fig. 8.5. Light dose response for photochemical transfection. HCT 116 cell were treated with $AlPcS_{2a}$ over night, washed for 4 h and incubated with a pEGFP-N1/polylysine complex for 4 h. The cells were then illuminated with red light, and 48 h later EGFP-expressing cells were scored by flow cytometry as described in the legend to Fig. 8.4. Cell survival was measured by the MTT assay as described earlier.[52] ■ – EGFP-positive cells; ● – cell survival. Error bars are SEM of three experiments.

cells. Thus, in this experiment an increase from 1% EGFP-positive cells at 0 min of light to 50% positives at 5 min of light was observed, representing a light-induced enhancement of transfection efficiency of about 50 times. Likewise, for HCT 116 cells there was a clear light dose response with maximum transfection being obtained with an illumination time of 6 min in this cell line (Fig. 8.5). In both these cell lines the photochemical enhancement of transfection was effective over a quite large range of light doses, a finding that may have important implications for the potential clinical use of the technology (see *PCI for Light Directed Gene Delivery in* In Vivo *Gene Therapy* section). It can also be seen (Fig. 8.5) that the transfection efficiency tends to go down at the very highest light dose, a phenomenon that has also been found in other cell lines. There are several possible explanations for this. First, it may be due to a more general cytotoxic effect of the photochemical treatment, e.g., destroying the cell's ability to transcribe or translate the foreign DNA. Second, it may be caused by photochemically induced oxidative damage to the transfecting DNA in the endocytic vesicles, e.g., generating inactivating mutations.

Photochemical Transfection in Different Cell Lines

Since most transfection methods are very dependent on the cell line, several different cell lines were tested for the effects of PCI on polylysine-mediated transfection. Because the optimal light dose will vary between different cell lines, the cells were compared at light doses killing about 50% of the cells, so that the photochemical effects should be the same in all cell lines. It can be seen (Fig. 8.6), that after light treatment the absolute level of transfection achieved varied between different cell lines, with a light-induced enhancement of between 32 (HCT 116) and 10 (THX) times in these experiments. In the WiDr colon carcinoma cell line virtually no transfection was observed even after PCI treatment, although we have shown in studies with an immunotoxin that PCI per se works well in this cell line.[55] So apparently in this cell line other barriers for transfection are more important than endosomal escape.

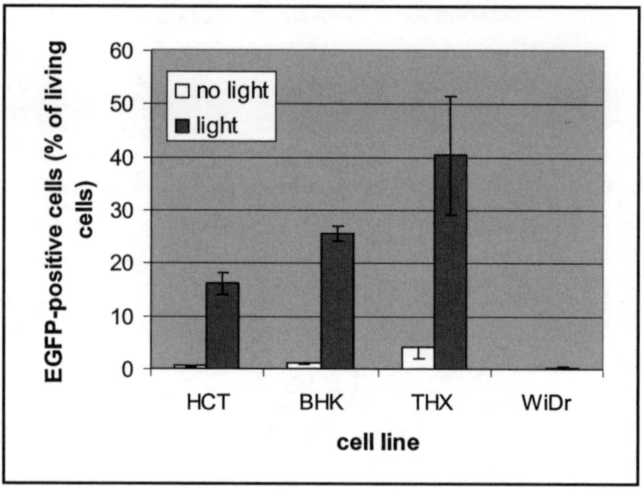

Fig. 8.6. PCI-induced transfection of different cell lines. Cells were grown, treated with $AlPcS_{2a}$ and incubated with pEGFP-N1/polylysine complexes (charge ratio 1.7) as described in the legend to Fig. 8.3. The cells were subjected to a light dose killing about 50% of the cells (4 min for THX, 8 min for HCT 116, 9 min for BHK-21, 5 min for WiDr) and analyzed for EGFP-expression by flow cytometry as described (52). Error bars represent interexperiment standard deviations of 16 experiments (THX), 3 experiments (HCT 116), and 2 experiments (BHK-21 and WiDr).

So far the conditions for PCI-induced transfection have only been partly optimized for polylysine-mediated transfection in THX-cells, where we have been able to get about 70% EGFP positive cells.[52] In the other cell lines tested, we have used the conditions optimal for THX-cells, but so far no optimization has been performed, except for determining the optimal light dose. However, in preliminary experiments with HCT 116 cells (M. Hellum, unpublished observation) transfection seems to increase both with the DNA dose and with the transfection time in a manner different from what was found for THX-cells,[52] indicating that conditions can be found where the transfection efficiency in these cells can be substantially increased as compared to what has been achieved so far.

The Requirement for Endocytosis

Our hypothesis for the mechanism behind PCI-induced transfection dictates that endocytosis should be obligatory for the photochemically enhancement of transfection. We have therefore performed a number of experiments to investigate the role of endocytosis and endocytic transport in photochemical transfection.

A possible explanation for the observed findings could be that the PCI-treatment increases the overall cellular uptake of the transfecting plasmid. However, Southern-blotting experiments indicated that plasmid uptake was not increased by PCI,[7] a conclusion also supported by flow cytometry analysis of the uptake of a fluorescently labeled oligonucleotide. Thus, these studies indicated that PCI does not induce a general increase in the plasma membrane permeability.[44] It was, however, still possible that PCI induces the transfer of a small subpopulation of plasmid molecules directly through the cell membrane and that this population gives rise to the observed PCI-induced transfection. However, several lines of evidence argue against this possibility. First, experiments at low temperatures where endocytosis is inactive, but where diffusion over the plasma membrane can still occur, showed that endocytosis was necessary for PCI-induced transfection.[7] Moreover,

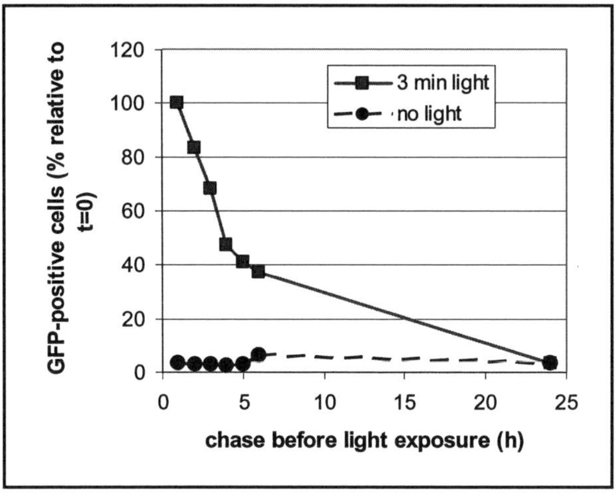

Fig. 8.7. Effect of chase time on the efficiency of photochemical transfection. $AlPcS_{2a}$-treated THX cells were incubated with pEGFP-N1/polylysine complex for 30 min (pulse) followed by chase in complex-free medium at 37°C for different time periods before 3-min light exposure. The amount of EGFP-expressing cells was calculated taking as 100% the percentage of EGFP-positive cells in light-treated cells at $t = 0$ (no chase). The experiment was designed so that the "washing time" in photosensitizer-free medium was the same for all samples, to ensure that the photochemical effects should be the same. Error bars are SEM of three experiments.

experiments with the inhibitor bafilomycin A indicated a requirement for endocytosis, since this inhibitor completely blocked the photochemical stimulation of transfection.[7] Bafilomycin A is supposed to block endosomal acidification and endocytic transport, but apparently does not effect uptake over the plasma membrane.[58,59]

The Southern-blotting experiments also indicated that PCI has no effect on the overall intracellular stability of the transfecting plasmid.[7] However, such studies must be interpreted with caution, since it is possible that the plasmid molecules giving rise to the observed transfection might be recruited from a small subpopulation of plasmid molecules whose behavior might be different from what is observed in studies on bulk plasmid DNA.

Since the PCI technology is intimately associated with the endocytic process, the relationship between the timing and length of the transfection period and the timing of the light treatment could be very important. For THX-cells it was always best to irradiate the cells directly after incubation with the DNA. If a chase period was included between incubation of the cells with DNA and the moment of irradiation, the effect of the light treatment decreased (Fig. 8.7).[7] This indicates that PCI works best when the plasmid is in early endocytic vesicles, a view also reflected in results showing that while the transfection efficiency increased sharply with time for transfection times below 1 h, only a modest increase could be observed by increasing the transfection time further.[52] The reason for these observations probably is that DNA in later endocytic vesicles has been degraded by endosomal or lysosomal nucleases, but it is also possible that the photochemical release per se is more effective from early that from late vesicles (Fig. 8.8).

Effects of PCI with Different Photosensitizers

It is inherent in our hypothesis for photochemical transfection that only photosensitizers localizing in endocytic vesicles should improve transfection. In accordance with this

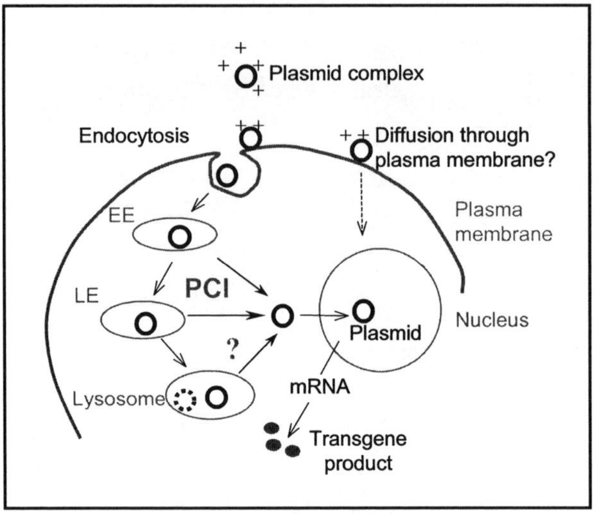

Fig. 8.8. Schematic representation of a plausible mechanism for PCI-mediated transfection. The positively charged plasmid complex binds to the cell surface by electrostatic interaction and is taken up into the cell by endocytosis (our data indicate that direct diffusion through the plasma membrane is unlikely, see text). The complex is transported along the endocytic pathway consisting of early endosomes (EE), late endosomes (LE), and lysosomes. PCI releases the plasmid from the endocytic vesicles. In principle PCI can release the plasmid from all such vesicles, however, the data indicate that release from endosomes is the most important (see text). After release from the vesicles the plasmid enters the nucleus, to be transcribed and translated into a protein product.

nonlysosomally localized photosensitizers like 3THPP (tetra(3-hydroxyphenyl)porphyrin) and 5-ALA (5-aminolevulinic acid)-induced PpIX (protoporphyrin IX) do not significantly stimulate polylysine-mediated transfection (Fig. 8.9). Furthermore, among photosensitizers localized in endocytic vesicles, differences in the transfection enhancing effects could also be observed, indicating that amphiphilic photosensitizers expected to localize mainly in the membranes of the endocytic vesicles (see *Intracellular Localization of the Photosensitizer* section) give the best effect.[44] Thus, localization of the photosensitizer in the membranes of endocytic vesicles seems to be important for optimal photochemical internalization of intact DNA molecules.

PCI with Different Transfection Agents

From Fig. 8.10 it can be seen that in THX cells PCI had a positive effect on transfection with polycations such as polylysine and polyethylenimine, while transfection with the cationic lipids DOTAP (*N*-[1-(2,3-dioleoyloxy)propyl]-*N*,*N*,*N*-trimethylammonium methyl-sulfate), Lipofectin, and βAE-DMRIE (β-aminoethyl-dimyristoyl Rosenthal inhibitor ether)[60] was actually reduced. Thus, apparently there is a difference in response between polycationic and cationic lipid transfection agents. This difference seems to be cell line dependent, since in HCT 116 cells preliminary experiments indicate that transfection mediated by cationic lipid transfection agents can be positively affected by PCI (M. Hellum, unpublished observations). The observed differences in the PCI response between the different transfection agents raise several interesting questions. It seems logical to think

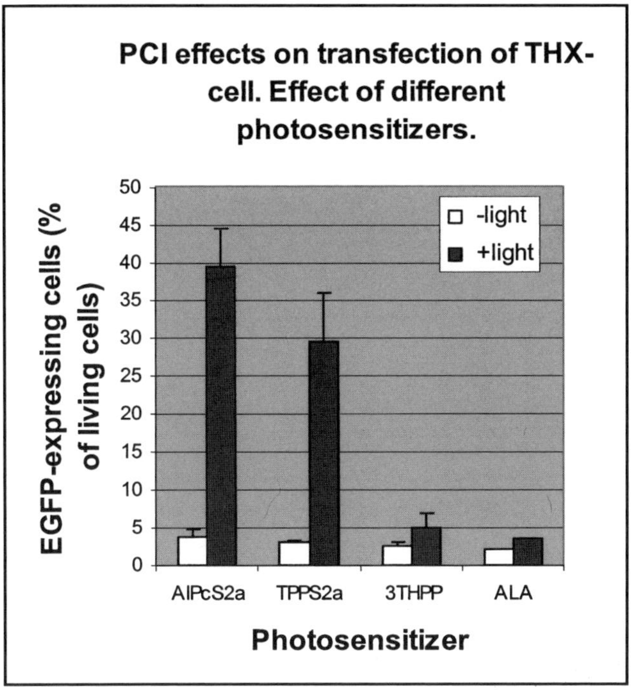

Fig. 8.9. Effect of photochemical treatment with different photosensitizers on transfection of THX cells. The cells were treated with different photosensitizers (20 µg/mL $AlPcS_{2a}$, 2 µg/mL $TPPS_{2a}$, 0.25 µg/mL 3THPP, or 1 mM ALA) before incubation with pEGFP-N1/polylysine for 4 h.[44] After transfer into fresh medium the cells were exposed to photosensitizer-activating light killing about 50% of the cells[44] and 2 days later the cells were analyzed for EGFP expression by flow cytometry.

that for transfection agents that are not positively affected by PCI, endosomal release is not a limiting factor for transfection. The *negative* effects seen with lipidic transfection agents might be due to photochemically induced toxicity having some generally inhibitory effect on transfection. If such effects are operative, they may also affect transfection by polycationic agents and may be reflected in the decrease in transfection observed at the higher light doses (see *Light Dose Dependence* section). If this is the case, then PCI-induced transfection could be viewed as a balance between these general negative effects and the positive effects caused by increased endosomal release. Thus, in the case of transfection agents that are very ineffective in endosomal release, such as polylysine, the positive effects would dominate, while for agents more efficient in such release the negative effects would dominate. There are, however, also several other possible explanations for the negative effect on lipoplex transfection. First, it is possible that the mode of endosomal release induced by PCI will only stimulate transfection for some transfection agents, while for others it might even have a negative effect. In the case of cationic lipid transfection agents, the model of Szoka and coworkers[31,32] suggests that dissociation of the lipid from the DNA occurs during endosomal release. A possibility is, therefore, that PCI will release the intact DNA/lipid complex, thereby bypassing an essential step of dissociation of the complex. It is also possible that DNA/lipid complexes are more prone to photochemically induced damage to the DNA than polycationic complexes, either because the lipids are

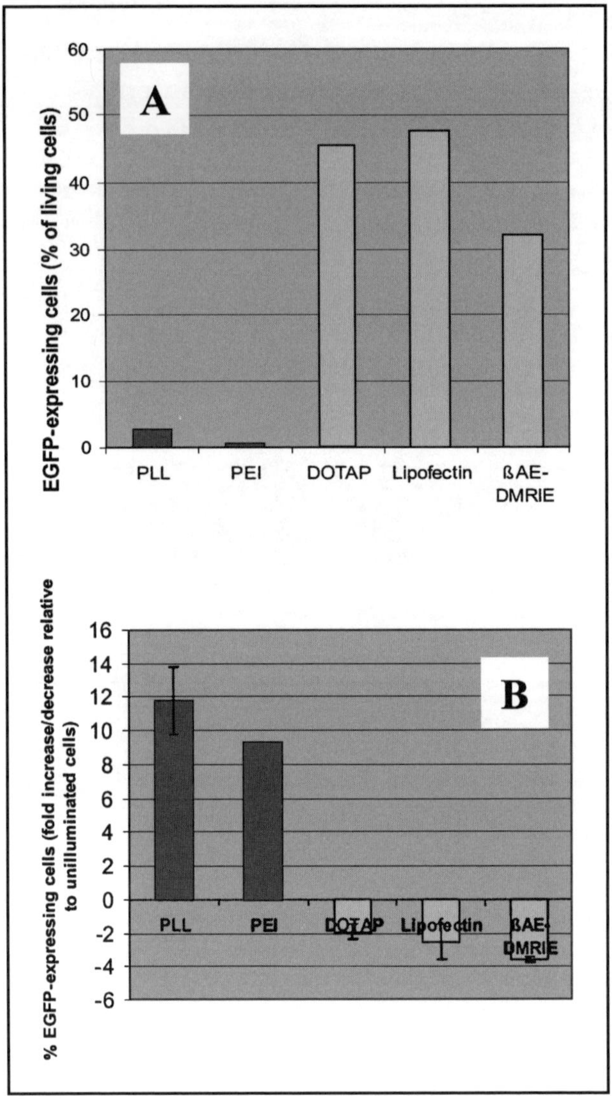

Fig. 8.10. Effect of photochemical treatment on transfection with different transfection agents. $AlPcS_{2a}$-treated THX cells were incubated with pEGFP-N1 plasmid complexed with different transfection agents for 4 h, transferred into fresh medium and exposed to a dose of red light killing about 50% of the cells (0.32 J/cm^2). Two days later EGFP expression was measured by flow cytometry. (A) EGFP expression in nonirradiated cells (% EGFP positive cells). (B) Fold increase/decrease in EGFP expression after light exposure, calculated taking the expression in nonirradiated cells as 1.

not effective in protecting the DNA against oxidative damage, or because lipid complexes are located closer to the photosensitizer and therefore are more liable to damage.

The observation that transfection by polyethylenimine (PEI) complexes is positively affected by PCI gives an indication that although PEI is often believed to induce endosomal escape; this process may not be very effective. As pointed out by Ogris et al.,[11] this may be related, e.g., to the size off the PEI/DNA complexes employed.

Targeted Vectors

To study whether PCI would also have positive effects on transfection when the DNA is delivered by receptor-mediated endocytosis, we studied transfection with transferrin–polylysine. In HCT 116 cells PCI had an even better effect on transfection mediated by transferrin–polylysine than on transfection with unmodified polylysine.[7] However, in THX cells, no transfection with transferrin–polylysine could be observed, neither with nor without PCI (not shown). Thus, PCI can also work very well with receptor-mediated transfection agents, but substantial cell line differences exist, maybe related to the level of receptor expression in the target cells.

Is Stimulation of Transcription Involved in PCI-Induced Transfection?

An aspect that deserves further consideration is the possibility that the photochemical treatment can stimulate transcription from the CMV (cytomegalovirus) promoter employed. Thus, it is well known that photochemical treatments induce cellular stress responses,[61,62] and it has been reported that at least some stress responses can stimulate transcription from the CMV promoter.[63,64] Moreover, the CMV promoter can be induced by ionizing and UV irradiation.[65,66] However, we have several reasons to believe that stress induction of the CMV promoter is not a major mechanism behind the stimulation of transfection by PCI: (i) Light-induced stimulation of transfection was not seen with non lysosomally localized photosensitizers at conditions inducing equivalent levels of cellular stress (Fig. 8.9). (ii) The stimulation is dependent on the type of transfection agent. One would not expect the CMV-promoter in the transgene to respond differently to stress-induction depending on the kind of transfection agent used for delivering the transgene. (iii) Illumination 24 h after delivery of the DNA does not stimulate transfection, neither polylysine- (Fig. 8.8) nor DOTAP-mediated,[7] in our system. At this time point one should expect the transfecting DNA to be located in the nucleus of the cell where transcription could readily be induced by stress. Thus, Bruening et al. (1998)[64] reported that stress applied up till at least 40 h after transfection could stimulate transcription from the CMV-promoter of transfected plasmids, and the same timing should be expected after PCI-induced transfection if such stimulation was the main mechanism behind the stimulation. (iv) In experiments on cells containing a chromosomally integrated EGFP-gene controlled by the CMV-promoter no increase in the level of EGFP-expression could be observed when the cells were subjected to $AlPcS_{2a}$ photochemical treatment (data not shown), arguing against a photochemically induced stimulation of transcription from the CMV-promoter under the conditions used for $AlPcS_{2a}$-mediated PCI.

However, it is still possible that stress-induced stimulation of transcription could play a role in the light-induced effects on transfection. Especially, there are ample possibilities for additive or counteracting effects where a stimulation of transcription could act together with other factors to determine the end result on transfection. Thus, the overall PCI-effects on transfection could be a combination of effects both on endosomal release and on transcription, and the mechanisms leading to the observed effects may be different between different combinations of cells, photosensitizers and transfection agents.

Discussion and Future Perspectives

PCI can potentially be exploited for several different purposes. Probably the most intriguing possibility is to use the technology as a method for site-directed enhancement of gene delivery in in vivo gene therapy. However, the technology should also be useful as an in vitro transfection method and as a method for investigating the importance of various cellular events and compartments in the process of transfection.

PCI for in Vitro Transfection

PCI seems to be able to induce transfection in a variety of different cell lines, and at least in some cases resulting in transfection efficiencies greater than those which can be obtained by other transfection methods (M. Hellum, unpublished observations). This indicates that PCI may be a useful method for transfecting cells in vitro, adding to the existing arsenal of laboratory gene transfer methods. Especially, it will be interesting to see if PCI can stimulate transfection in cells that are "notoriously" difficult to transfect with existing methods. Although the light-induced toxic effect is an apparent drawback for PCI-induced transfection, toxicity is also an often-encountered problem for most other efficient transfection methods. Thus, in the end the toxic effects associated with PCI might well be offset by the ability to use lower amounts of potentially toxic transfection agents due to the increase in transfection efficiency achievable by PCI. Furthermore, PCI as a transfection method is as yet far from being optimized, and it is reasonable to think that conditions can be found where the toxicity can be reduced while maintaining the positive effects on transfection.

PCI as a Tool for Investigating Cellular Mechanisms of Transfection

PCI could also be used as a method for analyzing the molecular mechanisms of transfection, especially in examining the importance of barriers of transfection, including endosomal escape, cytoplasmatic stability of the transgene, and nuclear uptake. The instant effect of the light treatment is especially useful and should make PCI a very valuable tool for tracking down the timing of several critical events in the process of transfection.

PCI for Light-Directed Gene Delivery in in Vivo Gene Therapy

As discussed by many authors (see, e.g.,[2,67] a long sought-after goal in in vivo gene therapy is to be able to specifically deliver a therapeutic gene so that it will be efficiently expressed at a desired location in the body but not at other sites. With a light-directed enhancement of gene delivery, the PCI technology may contribute to reaching this goal. In vitro PCI has shown an effect on transfection in several different cell lines, with polyplex transfection agents usually giving an increase in the number of transfected cells of about 20 times (Figs. 8.4–8.6). Moreover, in some cell lines there is also an increase in the transgene expression among the positive cells (measured as mean EGFP fluorescence in these cells), so that the total induction of transgene expression is more than 20-fold, e.g., in HCT 116 cells it can be in the order of 100-fold (data not shown). However, this effect seems to be dependent on the cell line, e.g., in THX cells no effect on the mean fluorescence of the positive cells was observed.[52]

It is, of course, a crucial question whether the effects seen in vitro would also be present in an in vivo situation. From animal studies with the protein toxin gelonin, we know that photochemical internalization per se works very well in vivo. Thus, in a model where $AlPcS_{2a}$ was administered intraperitoneally followed by intratumoral injection of gelonin, light treatment of the tumor showed that PCI dramatically and specifically increased the cytotoxic effects of gelonin.[57] Similar experiments combining PCI with gene delivery vectors are in progress in our laboratory.

One of the main obstacles for efficient gene delivery in vivo is that gene complexes are distributed very poorly through the target tissue, resulting in large areas of the target region receiving too low concentrations of gene complex to be efficiently transfected. In such a setting the enhanced transfection efficiency obtainable by PCI may be very valuable, making it possible to allow areas of poor complex penetration to be efficiently transfected. In addition, because of the light-directed nature of the technology, this could also be obtained without increasing transfection in nontarget areas, thus avoiding undesirable side effects. For cancer treatment there is also another possible element of specificity inherent in the PCI technology. Since photosensitizers in many cases tend to accumulate preferentially in tumor areas (see *Photochemically Induced Cytotoxicity and Photodynamic Therapy of Cancer* section), surrounding normal tissue may to a large extent be spared from the effects of the light treatment due to lower photosensitizer concentrations in the normal tissue. For example in brain tumors this effect could be very important since tumor:normal ratios in the photosensitizer concentration of >20:1 has been observed.[68]

The PCI-mediated specificity of transfection could also be increased by combination with other means for obtaining specific delivery and/or expression of the therapeutic gene. Thus, most methods for targeting the delivery of therapeutic genes[69–72] could probably be combined with PCI. Indeed, the employment of PCI with the targeted transfection agent transferrin–polylysine shows that PCI works very well also with a transfection agent taken up by receptor-mediated endocytosis.[7] Also transcriptional targeting[73,74] should be very well suited for combination with PCI. In addition to using promoters that will confer tissue specific expression, a very interesting approach would be to use promoters inducible by the photochemical treatment[75] to add further to the specificity achievable with the light treatment.

So far PCI has worked best with polycationic transfection agents such as polylysine and polyethylenimine, and the technology has generally worked best under conditions where the transfection efficiency of nonirradiated cells was low (Fig. 8.10). This is an advantage if the aim is to obtain maximum specificity by the light treatment, but it may be a disadvantage from the point of view of maximizing the overall transfection of the target cells. As discussed above in our first experiments with lipoplex transfection agents, PCI had a negative effect on transfection. However, preliminary experiments seem to indicate that conditions can also be found where lipoplex transfection can be enhanced even in cases where there is also substantial transfection of nonirradiated cells (M. Hellum, unpublished observations). This indicates that after proper optimization, PCI can substantially improve transfection also with transfection agents that on their own give a high transfection efficiency, which of course could be very important, both for in vivo and for in vitro applications.

The two main drawbacks for using PCI in in vivo gene therapy probably are the limited penetration of light into the tissues, and the cytotoxic effect of the photochemical treatment. In photodynamic therapy the range of the treatment usually is limited to a few millimeter from the tissue surface, depending among other things on the photosensitizer used. For PCI the reach would be expected to be greater, since transfection can be substantially induced also at relatively nontoxic light doses (Fig. 8.5). Thus, our in vivo experiments with gelonin (see above) show that tumors up to a thickness of at least 6 mm may be totally eradicated by PCI treatment, indicating an effect of the light treatment up to at least 6 mm from the tissue surface.

Light-induced toxic effects are an obvious drawback for some, but by no means all, in vivo applications. For gene therapy approaches where it is essential that the cells in the target area are not harmed, these effects generally cannot be tolerated. However, in many clinical situations, e.g., in cancer gene therapy, a certain amount of cell killing may not be harmful, or may even be beneficial. Furthermore, in cancer therapy the PCI treatment may, in addition to killing tumor cells, also induce the production of immune stimulating factors such as cytokines or heat-shock proteins,[40,76–79] that may help in raising a potent immune response to the cancer cells.[80,81] Although the photochemical treatment induces cytotoxic effects, a very important point is that these are generally restricted to illuminated areas of the body.[82] The photosensitizers in themselves usually have very little systemic toxicity, in contrast to other agents (e.g., chloroquine) effective in increasing endosomal escape. Thus, damage to vital organs could generally be avoided. Furthermore, the gain in efficiency obtainable by PCI may make it possible to use lower doses of the gene therapeutic agent, thus decreasing possible adverse effects of both the gene delivery vehicle and the expression of a therapeutic gene in nontarget cells. Also, there is substantial clinical experience with relevant photosensitizers, showing that they can be used safely in humans.[40] Work is in progress to examine PCI as a gene delivery method in vivo, as well as to further explore the use of PCI with different transfection agents for in vitro/ex vivo transfection.

Literature Cited

1. Anderson WF. Nature 1998; 392:25-30.
2. Verma IM, Somia N. Nature 1997; 389:239-242.
3. Greber UF, Willetts M, Webster P et al. Cell 1993; 75:477-486.
4. Leopold PL, Ferris B, Grinberg I et al. Hum Gene Ther 1998; 9:367-378.
5. Zabner J, Fasbender AJ, Moninger T et al. J Biol Chem 1995; 270:18997-19007.
6. Tseng WC, Haselton FR, Giorgio TD. J Biol Chem 1997; 272:25641-25647.
7. Prasmickaite L, Høgset A, Tjelle TE et al. J Gene Med 2000; 2:477-488.
8. Wilke M, Fortunati E, van den BM et al. Gene Ther 1996; 3:1133-1142.
9. Pouton CW. Adv Drug Deliv Rev 1998; 34:51-64.
10. Luo D, Saltzman WM. Nat Biotechnol 2000; 18:33-37.
11. Ogris M, Steinlein P, Kursa M et al. Gene Ther 1998; 5:1425-1433.
12. Pouton CW, Lucas P, Thomas BJ et al. J Control Release 1998; 53:289-299.
13. De Smedt SC, Demeester J, Hennink WE. Pharm Res 2000; 17:113-126.
14. Zhou X, Huang L. Biochim Biophys Acta 1994; 1189:195-203.
15. Zelphati O, Szoka FC, Jr. Pharm Res 1996; 13:1367-1372.
16. Pouton CW, Seymour LW. Adv Drug Deliv Rev 1998; 34:3-19.
17. Cotten M, Langle-Rouault F, Kirlappos H et al. Proc Natl Acad Sci USA 1990; 87:4033-4037.
18. Wagner E, Zatloukal K, Cotten M et al. Proc Natl Acad Sci USA 1992; 89:6099-6103.
19. Midoux P, Mendes C, Legrand A et al. Nucleic Acids Res 1993; 21:871-878.
20. Wolfert MA, Seymour LW. Gene Ther 1998; 5:409-414.
21. Erbacher P, Roche AC, Monsigny M et al. Exp Cell Res 1996; 225:186-194.
22. Curiel DT, Agarwal S, Wagner E et al. Proc Natl Acad Sci USA 1991; 88:8850-8854.
23. Cristiano RJ, Smith LC, Woo SL. Proc Natl Acad Sci USA 1993; 90:2122-2126.
24. Harbottle RP, Cooper RG, Hart SL et al. Hum Gene Ther 1998; 9:1037-1047.
25. Zauner W, Kichler A, Schmidt W et al. BioTechniques 1996; 20:905-913.
26. Zauner W, Kichler A, Schmidt W et al. Exp Cell Res 1997; 232:137-145.
27. Felgner JH, Kumar R, Sridhar CN et al. J Biol Chem 1994; 269:2550-2561.
28. Farhood H, Serbina N, Huang L. Biochim Biophys Acta 1995; 1235:289-295.
29. Wrobel I, Collins D. Biochim Biophys Acta 1995; 1235:296-304.
30. Boussif O, Lezoualc'h F, Zanta MA et al. Proc Natl Acad Sci USA 1995; 92:7297-7301.
31. Zelphati O, Szoka FC, Jr. Proc Natl Acad Sci USA 1996; 93:11493-11498.
32. Xu Y, Szoka FC, Jr. Biochemistry 1996; 35:5616-5623.
33. Berg K, Selbo PK, Prasmickaite L et al. Cancer Res 1999; 59:1180-1183.
34. Moan J, Sommer S. Cancer Res 1985; 45:1608-1610.
35. Weishaupt KR, Gomer CJ, Dougherty TJ. Cancer Res 1976; 36:2326-2329.
36. Jori G, Spikes JD. In Topics in Photomedicine, KC Smith (ed.), Plenum Press, New York, 1984, pp.183-318.
37. Moan J, Berg K. Photochem Photobiol 1991; 53:549-553.
38. Pass HI. J Natl Cancer Inst 1993; 85:443-456.

39. Henderson B, Dougherty TJ. Photochem Photobiol 1992; 55:145-157.
40. Dougherty TJ, Gomer CJ, Henderson BW et al. J Natl Cancer Inst 1998; 90:889-905.
41. Moan J, Berg K, Kvam E et al. CIBA Found Symp 1989; 146:95-107.
42. Berg K, Bommer JC, Winkelman JW et al. Photochem Photobiol 1990; 52:775-781.
43. Berg K, Western A, Bommer JC et al. Photochem Photobiol 1990; 52:481-487.
44. Prasmickaite L, Høgset A, Berg K. Photochem Photobiol 2001; 73:388-395.
45. Berg K, Moan J. Int J Cancer 1994; 59:814-822.
46. Peng Q, Nesland JM, Madslien K et al. SPIE 1996; 2628:102-113.
47. Moan J, Berg K, Bommer JC et al. Photochem Photobiol 1992; 56:171-175.
48. Maman N, Dhami S, Phillips D et al. Biochim Biophys Acta 1999; 1420:168-178.
49. Moan J, Berg K, Anholt H et al. Int J Cancer 1994; 58:865-870.
50. Wilson PD, Firestone RA, Lenard J. J Cell Biol 1987; 104:1223-1229.
51. Miller DK, Griffiths E, Lenard J et al. J Cell Biol 1983; 97:1841-1851.
52. Høgset A, Prasmickaite L, Tjelle TE et al. Hum Gene Ther 2000; 11:869-880.
53. Berg K, Sandvig K, Moan J. Patent 1996; PCT/NO95/00149.
54. Prasmickaite L, Høgset A, Berg K. In Methods in Molecular Medicine, Vol. 69: Gene Therapy Protocols, 2nd edn. JR Morgan (ed.), Humana Press, Totowa, 2000; pp. 123-135.
55. Selbo PK, Sivam G, Fodstad Ø et al. Int J Cancer 2000; 87:853-859.
56. Selbo PK, Sandvig K, Kirveliene V et al. Biochim Biophys Acta 2000; 1475:307-313.
57. Selbo PK, Sivam G, Fodstad Ø et al. Int J Cancer 2001; 92:761-766.
58. Clague MJ, Urbé S, Aniento F et al. J Biol Chem 1994; 269:21-24.
59. van Deurs B, Holm PK, Sandvig K. Eur J Cell Biol 1996; 69:343-350.
60. Wheeler CJ, Sukhu L, Yang G et al. Biochim Biophys Acta 1996; 1280:1-11.
61. Luna MC, Wong S, Gomer CJ. Cancer Res 1994; 54:1374-1380.
62. Ryter SW, Gomer CJ. Photochem Photobiol 1993; 58:753-756.
63. Boom R, Sol CJ, Minnaar RP et al. J Gen Virol 1988; 69:1179-1193.
64. Bruening W, Giasson B, Mushynski W et al. Nucleic Acids Res 1998; 26:486-489.
65. Tang D-C, Jennelle RS, Shi Z et al. Hum Gene Ther 1997; 8:2117-2124.
66. Francis MA, Rainbow AJ. Photochem Photobiol 2000; 72:554-561.
67. Curiel DT. Ann NY Acad Sci 1999; 886:158-171.
68. Chopp M, Madigan L, Dereski M et al. Photochem Photobiol 1996; 64:707-711.
69. Dachs GU, Dougherty GJ, Stratford IJ et al. Oncol Res 1997; 9:313-325.
70. Zauner W, Ogris M, Wagner E. Adv Drug Deliv Rev 1998; 30:97-113.
71. Godbey WT, Wu KK, Mikos AG. J Control Release 1999; 60:149-160.
72. Sudimack J, Lee RJ. Adv Drug Deliv Rev 2000; 41:147-162.
73. Miller N, Whelan J. Hum Gene Ther 1997; 8:803-815.
74. Nettelbeck DM, Jérôme V, Müller R. Trends Genet 2000; 16:174-181.
75. Luna MC, Ferrario A, Wong S et al. Cancer Res 2000; 60:1637-1644.
76. Evans S, Matthews W, Perry R et al. J Natl Cancer Inst 1990; 82:34-39.
77. Kick G, Messer G, Goetz A et al. Cancer Res 1995; 55:2373-2379.
78. Gomer CJ, Ryter SW, Ferrario A et al. Cancer Res 1996; 56:2355-2360.
79. de Vree WJ, Essers MC, Koster JF et al. Cancer Res 1997; 57:2555-2558.
80. Todryk S, Melcher AA, Hardwick N et al. J Immunol 1999; 163:1398-1408.
81. Todryk SM, Melcher AA, Dalgleish AG et al. Immunology 2000; 99:334-337.
82. Moan J, Berg K. Photochem Photobiol 1992; 55:931-948.

Chapter 9

Sonoporation: Mechanical DNA Delivery by Ultrasonic Cavitation

Douglas L. Miller, Sorin V. Pislaru, and James F. Greenleaf

Abstract

Development of nonviral gene transfer methods would be a valuable addition to the gene-therapy armamentarium, particularly for localized targeting of specific tissue volumes. Ultrasound can produce a variety of nonthermal bioeffects via acoustic cavitation including DNA delivery. Cavitation bubbles may induce cell death or transient membrane permeabilization (sonoporation) on a single cell level, as well as microvascular hemorrhage and disruption of tissue structure. Application of sonoporation for gene delivery to cells requires control of cavitation activity. Many studies have been performed using in vitro exposure systems, for which cavitation is virtually ubiquitous. In vivo, cavitation initiation and control is more difficult, but can be enhanced by cavitation nucleation agents, such as an ultrasound contrast agent. Sonoporation and ultrasonically enhanced gene delivery has been reported for a wide range of conditions including low frequency sonication (kilohertz frequencies), lithotripter shockwaves, HIFU, and even diagnostic ultrasound (megahertz frequencies). In vitro, a variety of cell lines has been successfully transfected, with concomitant cell killing. In vivo, initial applications have been to cancer gene therapy, for which cell killing can be a useful simultaneous treatment, and to cardiovascular disease. The use of ultrasound for nonviral gene delivery has been demonstrated for a robust array of in vitro and mammalian systems, which provides a fundamental basis and strong promise for development of new gene therapy methods for clinical medicine.

Introduction

The role of genetic factors in the development of human disease is by no means a new concept. Genetic phenomena were already described by ancient Greeks and Hebrews. The nineteenth century brought the pioneering work of Gregor Mendel, who laid the grounds of modern genetics, and in 1902 Garrod described alkaptonuria as the first "inborn error of metabolism." However, it was only during the last two decades that our understanding

---

Abbreviations: CAT, chloramphenicol acetyltransferase; DNA, deoxyribonucleic acid; ELISA, enzyme-linked immunosorbant assay; FITC, fluoroisothiocyanate; GFP, green fluorescent protein; HIFU, high-intensity-focused ultrasound; PESDA, perfluorocarbon-exposed sonicated dextrose albumin; PRF, pulse repetition frequency; SW, shock wave.

Synthetic DNA Delivery Systems, edited by Dan Luo and W. Mark Saltzman.

of genetic mechanisms and associated technology have advanced enough to attempt therapeutic genetic manipulations. The newcomer in the therapeutic armamentarium, gene therapy, can be defined as the treatment of human disease by introduction of genetic material into human tissues. The transferred DNA can replace a defective gene, enhance, or inhibit a preexisting cell function, or introduce a completely new function into the cell. Conceptually, gene therapy could be used in a vast number of human diseases, beginning with the inborn errors of metabolism and ending with cancer therapy. It can be envisaged that, with the completion of the Human Genome Project, the list of diseases that are amenable to genetic therapy will only increase.

Identification of genetic mechanisms responsible for disease and the design of new genetic strategies alone cannot guarantee efficient gene therapy. The necessary step in all forms of genetic manipulation is transfection, the introduction of genetic material into cells. This is a major limiting step because DNA is a large, highly charged molecule, and, therefore, cannot diffuse through the cell membrane. Once inside the cell, the DNA is expressed for various periods of time. In this sense, we could distinguish stable transfection (achievable by incorporation of the transferred gene into the genome) and transient transfection (when the genetic fragment is expressed, but not included in the chromosomes). Strategies for achieving both types of transfection exist, but current techniques pose most of the limitations to the clinical application of gene therapy. Until now, none of the methods tested in the clinical setting have allowed an effective, safe, minimally invasive delivery of the genetic material.

Transfection protocols can be divided into two broad categories—viral and nonviral.[1] The major vectors used so far in clinical testing are adenoviral (for cardiovascular disease) and retroviral (especially for cancer therapy); adeno-associated and lentiviral vectors are currently under investigation. Adenoviral vectors are engineered by removing key sequences for replication from the viral genome (thus rendering the virus incapable of replication) and replacing them with the therapeutic gene.[2] Once injected into the organism, the adenoviral vector binds to membrane receptors and is subsequently internalized. The substituted DNA is transferred to the nucleus where it remains episomal (is not included in the host cell chromosomes). Thus, transfection with adenoviral vectors is only transient but, on the other hand, poses no concern for mutagenesis. The major advantages of this vector are related to its ability to transfect nonreplicating cells and the high percentage rates of transfection. Retroviral vectors are restricted to cells that proliferate actively, but the genetic material is incorporated into the host genome, allowing stable transfection. Although viral-mediated transfection allows high transfection rates when compared to DNA alone, currently available viral vectors still have substantial drawbacks. The major concerns are related to potential toxicity, development of severe immunological reactions (for adenoviruses), and potential induction of new mutations (for retroviruses).

Forms of nonviral transfection are represented by lipofection, electroporation, particle bombardment, and sonoporation. Lipofection is one of the most widely used nonviral methods of transferring genetic material to living cells. Essentially, cationic lipids encapsulate the negatively charged DNA and facilitate transfer of the gene through the cell membrane.[3] This method allows high transfection rates with minimal cellular toxicity, but as with the viral vectors, it does not allow control of spatial or temporal specificity of delivery. Electroporation refers to the transfer of DNA through membrane pores formed in high-voltage electric fields.[4] This method allows some spatial targeting, but requires electrode placement, which can be invasive. Particle bombardment represents yet another way of "injecting" foreign DNA into cells, this time by coupling the gene to projectiles that are made to penetrate the membrane at high speed.[5] This method also allows accurate placement of DNA delivery, but appears to be limited to surface (e.g., skin) applications.

Sonoporation, the subject of this chapter and a recent review by Newman et al.,[6] is a relative newcomer in gene transfer. This technique is designed to enhance cell permeability through the use of ultrasound. Uptake of both DNA and other macromolecules has been demonstrated by several investigators.[7–11] However, research in the field gained momentum only after the identification of cavitation as the probable mechanism behind the increased cell permeability, and the demonstration of further enhancement of transfection efficiency by using cavitation nuclei, such as ultrasound contrast agents.[8,10,12,13] These findings opened tremendous opportunities for targeted gene transfer. Conceptually, gene vectors mixed with ultrasound contrast agents could be injected intravenously, and targeted gene transfer could be achieved by selective insonation of a predefined area. Indeed, promising results have been recently reported in animal models.[14–18] By using this approach, the risk of systemic exposure (a major drawback of current clinical protocols of gene transfer) could be substantially reduced.

There are, of course, other approaches to local gene therapy (eluting polymers, coated stents, local delivery catheters, direct injection, etc.). However, most of the local delivery systems tested so far provide a high local-to-systemic ratio at only one particular location (either the vascular segment that is treated or the injection site). This is of course an advantage when treating a very localized disease, such as restenosis after angioplasty. On the other hand, most of the pathological processes amenable to gene therapy, even when localized (myocardial ischemia, tumors, etc.), would require a diffuse treatment of the affected organ that can be achieved only by enhancing the transfer at the microcirculation level. This might be achieved by sonoporation, which combines the capability of enhancing gene transfer with the possibility of restricting this effect to the desired area, at the desired time (spatial and temporal specificity).

To better present the use of ultrasound for transfection, we will first describe basic ultrasonic principles, consider the range of potential bioeffects of ultrasound, and briefly outline uses of ultrasound in biomedicine. Research on sonoporation, which is the biophysical phenomenon behind the transient enhancement of cell permeability, will be examined. The current status of ultrasound-aided DNA transfer will then be reviewed for in vitro and in vivo conditions. Finally, potential clinical applications and future directions will be summarized.

Ultrasound and Ultrasonic Biological Effects

Ultrasound

Ultrasound propagates through matter from a vibrating object as an oscillating pressure wave. This is fundamentally the same as audible sound, but frequencies f are higher than 20 kHz, which is above the range of human hearing. The pressure amplitude of a sinusoidal wave is typically specified in Pascal (Pa, kPa, or MPa, where 100 kPa is approximately atmospheric pressure). The speed c of propagation is about 1500 m s^{-1} in aqueous media and the wavelength is defined by c/f, which is about 1.5 mm at 1 MHz. Wavelength is an important determinant of the geometry of a wave propagating away from a vibrating object: objects smaller than a wavelength in extent yield spherically diverging waves, while relatively large objects can yield directional beams and focusing. The propagating wave transmits energy, and exposure is often specified in terms of the intensity (e.g., W cm^{-2}), which is proportional to the square of the pressure amplitude.

As an ultrasound wave propagates, several phenomena occur, which can potentially cause physical perturbations of biological cells and tissue. The primary mechanisms for bioeffects of ultrasound are heating and cavitation. Heating occurs as the ultrasonic energy is absorbed in a medium. At 1 MHz, absorption in tissue can be quite important in biomedical applications, and at 1 MHz, for example, intensity can by halved for each 10 cm of propagation (i.e., for a typical 0.3 dB cm^{-1} MHz^{-1} absorption coefficient). The lost intensity appears mostly as heating of the tissue, which is normally more important for higher frequencies. The amount of heating depends not only upon generating mechanisms, but also on the geometry of energy deposition, and on the heat transfer properties of tissues via conduction, convection, and radiation. The picture is further complicated by the generation of heat during cavitation and by the heating of the ultrasonic probe itself.

Cavitation is a nonthermal interaction between a propagating pressure wave and a gaseous inclusion in aqueous media. For low-pressure amplitudes, a gas bubble pulsates at amplitudes that depend strongly on its size, and are maximized at the resonance frequency of the bubble. At 1 MHz, a resonance size free bubble in water is about 7 μ in diameter, with cavitation phenomena typically being more important at lower frequencies. Large bubbles or air–water interfaces simply act as reflective surfaces for ultrasound waves. At higher pressure amplitudes, the transformation of cavitation nuclei, which are normally inactive and difficult to detect, into active cavities and bubbles gives the appearance of a threshold for vigorous cavitation. In addition, proliferation of bubble populations can multiply cavitation effectiveness. As a part of cavitation activity, microbubbles implode into minute sites of intense heat, light (sonoluminescence), free radical production, and shock wave generation, which can introduce secondary mechanisms for bioeffects.

Bioeffects of Ultrasound

The knowledge of bioeffects attributable to ultrasound extended remarkably during the past decades, as concerns regarding the safety of diagnostic and therapeutic ultrasound triggered research into the area (see Table 9.1). A detailed review of ultrasound bioeffects is beyond the scope of this chapter; the interested reader is referred to general information on bioeffects of ultrasound.[50–53]

The bioeffects of ultrasound generally may be categorized as thermal and nonthermal. Heating by more than a few degrees above normal biological temperatures can perturb biological systems; for example, by enhancing metabolism and perfusion of tissue. At high intensities, rapid heating can cause simple cooking of tissue. Generation of heat by ultrasound has several applications in medicine and biology, from cleaning tanks to induction of hyperthermia for surgery and tumor ablation, but heating may also be associated with deleterious effects, especially in tissues undergoing organogenesis or with poor heat-transfer mechanisms.[54] However, it is generally assumed that thermal mechanisms play only a minor role (if any) in the ultrasound-induced increase in cell permeability. A detailed review of thermal effects of ultrasound is provided by the 1992 report of the National Council on Radiation Protection and Measurements.[51]

Any process that can produce a biological effect without a significant degree of heating (i.e., less than about 1°C above physiological temperature) is a nonthermal mechanism. In the case of ultrasound, nonthermal bioeffects are coined "mechanical." However, one should keep in mind that the term "mechanical effect" includes also processes that are not of mechanical nature, but arise secondary to mechanical interaction between ultrasound and tissues, such as chemical reactions initiated by free oxygen species generated during cavitation and sonoluminescence.

The concentration of ultrasonic energy by cavitation bubbles yields a potential for biological effects in their vicinity whenever bubbles or cavitation nuclei are present in a biological medium exposed to ultrasound. Cavitation bioeffects can be confusing and

Table 9.1. Selected mechanical bioeffects of ultrasound

	Effects	Comments	Refs.
Cellular level			
Free-radical generation		Intracellular event?	(19)
Permeability	Increased ion conductance		(20, 21)
	Increased permeability to macromolecules		See Sonoporation section
Excitability	Enhancement and suppression		(22)
Agglutination	Erythrocyte agglutination		(23)
Growth	Transient decrease	Delayed cell lysis?	(24)
DNA	Single-strand break		(25)
	Increased sister chromatid exchange	Unconfirmed	(26, 27)
	Mutation frequency		(28)
Cell death			See Sonoporation section
Systemic level			
Cardiac	Premature contractions, decreased aortic pressure		(29, 30)
Capillary	Petechiae	Diagnostic ultrasound with contrast agent	(31, 32)
Vascular	Vasoconstriction, vessel rupture		(33)
Lung	Microvascular hemorrhage		(34–36)
Intestine	Microvascular hemorrhage		(37, 38)
Fetal development	Abnormal head development	Thermal mechanism Only at high pressure	(39)
Neural	Hearing and tactile perception	Radiation pressure effect	(38, 40)
Therapeutic effects			
Thrombolytic	Enhanced thrombolysis		(41, 42)
Antitumoral	Enhanced effect of cytotoxic agents		(43, 44)
Drug delivery	Local drug delivery		(11, 45, 46)
Gene therapy	Enhancement of gene transfer		See Lysis and Sonoporation
Ultrasonic surgery			(47)
Healing	Stimulated bone growth		(48, 49)

elusive to study owing to the requirement of preexisting gaseous inclusions and to difficult-to-predict cavitation thresholds. In some in vitro situations, cavitation readily occurs at modest pressure amplitudes, because of the presence of cavitation nuclei or suitably sized bubbles. In contrast, in vivo conditions in mammals typically minimize cavitation nuclei (owing to active filtering and sterilization by physiological processes), with the notable exceptions of the lungs and intestine. Cavitation causes mechanical perturbation in the vicinity of active bubbles, which can lead to membrane effects on individual cells, and to capillary rupture in vivo. The membrane effects range from sonoporation, which is the transient opening of holes in the membrane, to cell lysis or even fragmentation. Sonoporation is the effect of interest in this chapter, because cells exchange molecules with the surrounding medium while the transient holes are open and then survive for further study.

As could be expected, there is a strong influence of tissue characteristics on the bioeffects of ultrasound at any given level of insonation. With respect to mechanical bioeffects, the most important factor seems to be the presence or absence of natural gas bodies. In addition differences between species and differences related to the developmental age have been described.[52]

It can be inferred that tissues naturally containing gas–liquid interfaces would be more susceptible to mechanical effects because cavitation depends on the presence of gas bodies. Indeed, microscopic hemorrhage in the lungs and intestine was consistently demonstrated after ultrasound exposure across species.[34–37,55] Several mechanisms have been proposed to explain lung hemorrhage: cavitation in the gas-rich environment in the capillaries adjacent to the alveoli, migration of pockets of air from the lung parenchyma into the pleural fluid and pleural damage upon insonation, shearing of the air–tissue interface, and a variation of the spallation effect (i.e., rapid ejection of liquid into air when a shock wave hits the air–liquid interface).[52]

The occurrence of mechanical effects in tissues that do not naturally contain gas bodies is more difficult to explain. It is well known that cavitation is very difficult to induce in vivo, basically because the filtration system of mammalian organisms allows the presence of cavitation nuclei in only minute amounts, and only in particular sites. This observation concurs with the finding that the cavitation threshold for water doubles upon filtration to 2 μm.[56] However, it was hypothesized that in animals, bubbles can form spontaneously in low-surface tension fluids or by exclusion of water from hydrophobic surfaces.

Most of the studies performed with insonation within the low megahertz range (thermal effects are minimized at these frequencies) showed the presence of a threshold for the occurrence of mechanical bioeffects, that approximately coincides with the threshold for cavitation.[27] Parameters characterizing ultrasound obviously play a role in the generation of biological effects. Beyond the intuitive role of ultrasound wave intensity, exposure time, and central frequency, more subtle parameters such as type of application (continuous vs. pulsed), pulse repetition frequency (PRF), and duty cycle also influence the ultrasound effects. Clarke and Hill[57] were among the first to study the complex interplay between intensity, mode of exposure, and duty cycle. The threshold relationship between intensity and bioeffects observed with continuous wave exposure held true also when pulsed waves were used. However, the PRF and the duty cycle also played a role in the extent of effects generated.

Biomedical Applications of Ultrasound

Ultrasound and its bioeffects have been studied and exploited in biomedical applications for over 70 years. In diagnostic ultrasound imaging, frequencies from about 2 to 20 MHz are used under conditions of minimal bioeffects. Continuous or pulsed wave beams can detect blood or other motion by the Doppler effect. Pulse-echo systems with carefully focused and scanned beams of ultrasound are utilized for medical imaging in obstetrics, cardiology, and radiology. Typically, pulsed ultrasound has relatively low-time average intensities, leading to relatively low heating. Although the pulses can have relatively high pressure amplitudes (greater than 2 MPa), the relative lack of cavitation nuclei in vivo leads to relatively low risk of cavitation. A possible exception to this is the recent introduction of contrast agents into diagnostic ultrasound. These agents consist of suspensions of stabilized gas bodies of diameters suitable for passage through the circulation and also for return of strong echoes. Diagnostic ultrasound with contrast agents, therefore, introduces cavitation phenomena into an otherwise high threshold environment, with the potential for new therapeutic (or deleterious) bioeffects. For example, drug delivery to the interstitium can be accomplished by ultrasound-induced capillary rupture when microbubbles are present in the circulation.[45]

Therapeutic applications often exploit ultrasound-induced heating. This effect is used for physical therapy of muscle and joint disease, using broad-beam ultrasound applicators, and for surgery, using high-intensity-focused ultrasound (HIFU).[54] Nonthermal effects of ultrasound are also of value. Low-frequency (e.g., 20 kHz) ultrasound probes are used for cavitation-induced lysis and disruption of cells. Focused high amplitude shock waves are used for lithotripsy of kidney and gall bladder stones. These shock waves typically involve single cycle pulses of about 100 kHz ultrasound with negative pressures in excess of 5 MPa. Because of these high pressures, lithotripter shock waves can also cause cavitation in tissue, an unwanted side effect that may be exploited for therapeutic purposes. If cavitation nucleation can be controlled or enhanced in vivo, HIFU and even diagnostic ultrasound also have potential for nonthermal therapy via the cavitation mechanism. These exciting new possibilities for medical therapy have stimulated much recent research on sonoporation and DNA delivery with ultrasound.

Sonoporation

Membrane damage is a well-known biological effect of ultrasonic gas body activation or cavitation.[27,50] The mechanical action of the cavitation bubbles typically causes cell lysis and disintegration. However, sublethal membrane damage also occurs, in which large molecules in the surrounding medium are able to pass in or out of the cell, followed by membrane sealing and cell survival. This allows foreign macromolecules to be trapped inside the cell. This ultrasound-mediated increase in cell membrane permeability has been termed "sonoporation," a term that recognizes some similarity to electroporation. It should be noted that sonoporation represents transient permeabilization, which can be indicated by trapping large fluorescent molecules inside the viable cells (the molecules are excluded by viable cells and leak out of nonviable cells), and is different from the commonly noted permeabilization indicated by trypan blue or propidium iodide stains, which stain lysed, nonviable cells.

Sonoporation was initially demonstrated using 20 kHz sonication,[58] although this finding was not immediately noted as a new ultrasound bioeffect (it was intended as a method for intracellular pH measurement). Ameboid mold cells were sonicated in the presence of large fluorescent Dextran molecules. Cell recovery was about 40% of the original number, and 10% of these were loaded with the large molecules. Sonication treatment (20 kHz) has also been used to load mammalian cells with fluorescent Dextran.[59] Ten to 20% of sonicated hepatoma, mouse myeloma, HeLa, and fibroblast cell lines were found to be fluorescent after sonication with 10 mg/mL of 40 kD fluorescent Dextran. In another study using 20 kHz ultrasound, Johannes and Obe[60] showed that sonication of cultured CHO cells in the presence of endonuclease enzymes led to the production of chromosome aberrations in about 20% of subsequently cultured cells. Without sonication, few cells internalized the enzymes or had aberrations, indicating that the enzymes entered the cells via sonoporation.

Following the development of lithotripsy, research on lithotripsy-mode ultrasound quickly broadened to encompass a variety of biological effects.[61] An important factor found for shock wave treatments was that some cells were transiently permeabilized, allowing them to take up large molecules normally excluded by the cell membrane. In an in vitro study, Gambihler et al.[62] evaluated the accumulation of fluorescein-labeled Dextran (3900–2,000,000 molecular weight) in L1210 cells exposed in the presence of the Dextran. The large molecules, normally excluded by the cell membrane (except for pinocytosis into vacuoles), were found in surviving cells with a diffuse distribution within the cytoplasm. The sonoporation effect of lithotripter shock waves may be valuable for selectively treating tumors with large toxic molecules.[63] The ribosome inactivating proteins gelonin and saporin were apparently transferred into L1210 mouse leukemia cells and HELA cells by shock wave exposure, resulting in greatly enhanced cell killing. These plant toxins are also usable in vivo, and significant remission of SSK2 fibrosarcoma tumors was obtained in 40% of mice treated by systemic injection of the toxins and localized shock wave treatment.

Sonoporation of red blood cells was examined in relation to cavitation-induced hemolysis by Miller et al.[64] in whole blood, which simulates in vivo conditions but allows the convenient logistics of in vitro experiments. Fluorescent-labeled Dextran (FITC-dextran at 580,000 MW) was added to suspensions of canine erythrocytes, and the mixture was exposed to lithotripter shock waves of 14.8 MPa mean pressure amplitude. Exposure at 5 or 50% hematocrit in PBS or 50% in plasma yielded hemolysis and FITC-Dextran uptake in surviving cells. An air bubble was needed in the chamber to obtain substantial effects, implicating the cavitation mechanism. Hemolysis increased with increasing numbers of shock waves. The numbers of cells with fluorescent Dextran uptake remained roughly constant for 250–1000 shock waves, but this represented an increasing percentage of the surviving cells. These interesting exposure–response trends could be modeled by simple theory for random interaction of the cells with bubbles. The number of cells S, of the original number S_0, surviving n shock waves declines exponentially at a rate of a per shock wave, is given by

$$S = S_0\, e^{-an}. \tag{1}$$

The number of fluorescent cells F arising at a rate b from the surviving cells, and lysed at rate a, is then given by

$$F = S_0\, e^{-an}(1 - e^{-bn}). \tag{2}$$

This theoretical model has the interesting property that 100% of the survivors will eventually become loaded with fluorescent molecules as n increases toward infinity. This theory was fitted to actual measurements[64] as shown in Fig. 9.1. About one cell became fluorescent for every three cells lysed. The percentage of survivors that were fluorescent, plotted in the lower panel of Fig. 9.1, was calculated as F/S times 100%.

Sonoporation by shock waves has also been demonstrated by Zhong et al.[65] Mouse lymphoid cells were treated in the presence of fluorescent Dextran, and the percentage of transiently permeabilized cells reached 11.2% after 100 shock waves. The effects were associated with acoustic emissions indicative of cavitation. High-speed photography was used to show creation and collapse of bubbles with maximum sizes in the range of 100–200 μm, which emitted shock waves upon creation and collapsed up to 1 ms later.

The exact process by which cavitation causes transient permeabilization of cell membranes may be related to cavitation-generated stress waves. Laser pulses produce bubbles by

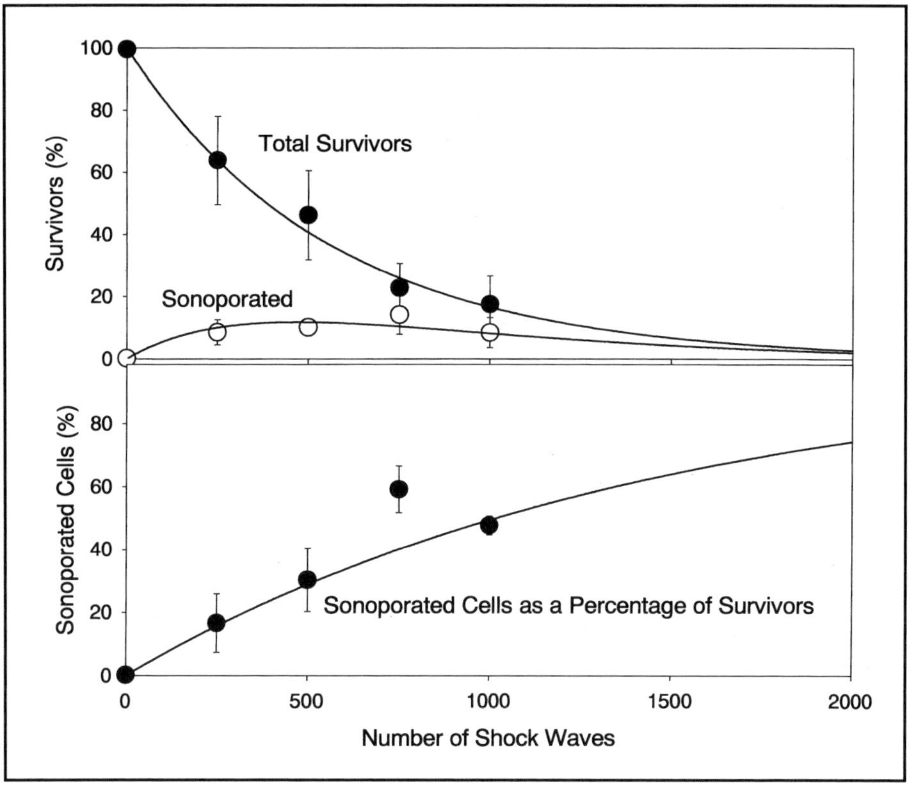

Fig. 9.1. The percentage of erythrocytes lysed and sonoporated (upper panel) for lithotripter shock wave treatment of whole blood with added fluorescent Dextran. The lower panel shows the percentage of sonoporated survivors. The plotted curves represent a theoretical model of the process, which was fitted to the data (see text). Reproduced with permission from Miller et al.[64]

optical breakdown of the medium, and the resulting generation of ultrasonic stress waves has been shown to transiently permeabilize cell membranes.[66] In studies of this effect in red blood cells (erythrocytes), an important role was demonstrated for aquaporins (water channels) in the membranes.[67] Human erythrocytes, which have aquaporins, showed uptake of fluorescent Dextran after stress wave treatment, while chicken red blood cells, which do not have aquaporins, did not. If the channels were blocked in human erythrocytes, the stress wave induced uptake of fluorescent Dextran was eliminated. Another important factor is the gradient (or rise time) of the stress waves.[68] Stress wave gradient was varied by applying ArF excimer or ruby laser pulses to polyamide or polystyrene targets, and transient permeabilization was observed by fluorescent Dextran uptake in cells resting on the target material. An 8-ns rise time was much more effective for cell loading than a 40-ns rise time for about the same peak stress (about 60 MPa). A single laser-generated stress wave was sufficient to achieve maximal cell loading of the large molecules.

A common exposure system used for in vitro research on cavitation bioeffects is the rotating tube system. Cells in suspension are affected by bubbles, which cycle back and forth across the tube as it rotates. Sonoporation of cultured cells has been demonstrated with this system using fluorescent Dextran as the test macromolecule.[8] Chinese hamster ovary cells were exposed to 2.25 MHz ultrasound in sterile 4.5-ml polyethylene chambers and tested for cell lysis, sonoporation, and DNA transfection. Ten percent of Albunex® (Mallinckrodt Inc., St. Louis, MO), a gas-body-based ultrasound contrast agent, was added to assure cavitation nucleation, and the chambers were rotated at 60 rpm to promote cavitation activity during the 1-min exposures. Sonoporation was observed for spatial peak pressure amplitudes as low as 0.1 MPa, and significant lysis occurred above 0.2 MPa. Up to half of the surviving cells were fluorescent after exposure at 0.8 MPa. The effects essentially disappeared when the Albunex® was omitted from the suspension and the tube was not rotated. The ultrasonic frequency has an important role in sonoporation.[69] After continuous wave exposure at frequencies of 1.0, 1.68, 2.25, 3.3, 5.3, and 7.15 MHz, sonoporation was noted for spatial peak pressure amplitudes as low as 0.1 MPa up to 3.3 MHz, increasing to 0.39 MPa at 7.15 MHz. Significant lysis occurred for 0.14 MPa exposures at 1.0 MHz, but not for the other frequencies with the relatively low-pressure amplitudes of exposure used. The fluorescent cell count increased roughly in proportion to increasing Albunex® concentration. The sonoporation effect appears to be a form of membrane damage, and although the cells exclude trypan blue dye immediately after exposure, they have poor proliferation ability in culture. The plating efficiency of cells exposed to 0.28 MPa at 2.25 MHz and sorted by a flow cytometer were only 19% (3.6% SD) for fluorescent cells, compared to 67% (1% SD) for non-fluorescent-exposed cells and 62% (6% SD) for sham-exposed cells.

Ward et al.[70] observed ultrasound induced cell lysis and sonoporation for 2 MHz, 0.2 MPa exposure of cell suspensions in a 200 rpm rotating tube system. Addition of ultrasound contrast agent was essential for robust effects. Optison™ (Mallinckrodt Inc., St. Louis, MO), an improved agent with greater persistence owing to the use of perfluoropropane gas, provided much greater enhancement of effects than did Albunex®, which contains air. In further work with Optison™,[71] the bubble to cell spacing appeared to have an important role in the magnitude of effects. The spacing was inferred from a static model (i.e., neglecting ultrasonic forces on the bubbles or between cells and bubbles) for different cell and bubble concentrations. The percentage of affected cells declined with concentration, which was approximately as the inverse third power of the static spacing.

Sonoporation has also been demonstrated for monolayer cells (i.e., a nonrotating system) in contact with Optison™ gas bodies for treatment by 3.5 MHz diagnostic ultrasound.[72] Epidermoid carcinoma cells were grown on thin Mylar sheets, which formed the upper window of a thin (1 mm) disc-shaped chamber. Fluorescent Dextran and Optison™ were added to the medium, and the gas bodies were allowed to rise to the cells before exposure. Up to about 10% of the cells in a 1-mm field of view were fluorescent, indicating sonoporation, after 1-min exposure in the spectral Doppler mode or two-dimensional scan mode. Pressure amplitudes of 0.23 MPa in Doppler mode (5 μs pulses) and 0.39 MPa (0.46 μs pulses) in scan mode were about equally effective. This finding indicates that even diagnostic ultrasound might be useful in therapeutic applications of sonoporation.

DNA Delivery by Sonoporation

The ability of ultrasound to load cells with large molecules, which survive for subsequent culture, opens the possibility of DNA transfection and expression of foreign gene products. The earliest work on this method was performed in vitro and in plant cells. Recent research has expanded to include the possibilities for gene therapy in mammalian cells by sonoporation in vivo.

DNA Delivery in Vitro

Sonication treatment (20 kHz) has been used to transfect cells with plasmid DNA.[59] Murine fibroblasts lacking the gene for the production of the enzyme thymidine kinase were used as the target cells for "therapy" with a plasmid (pPVTK4) coding for the enzyme. After sonication treatment in the presence of the plasmid, survival was about 70%. Transfection was demonstrated by culture with medium containing hypoxanthine, aminopterin, and thymidine (HAT) to select against colonies lacking the thymidine kinase gene. An average of 23 transformed colonies was counted per million viable cells, and repeated passage of selected cells indicated that some stable transfection had occurred.

Sonication of plant cells also results in loading of DNA plasmids into the cells. Joersbo and Brunstedt[73,74] found that plasmid DNA could be transferred into sugar beet and tobacco protoplasts by 20 kHz sonication as evidenced by chloramphenicol acetyltransferase marker gene expression. Furthermore, viruses could also be loaded into cells by this method.[75] Sugar beet protoplasts were sonicated with beet necrotic yellow vein virus particles. Infection was analyzed by ELISA assay of the virus protein coat and was maximal about 3 days after inoculation. Zhang et al.[76] reported transient expression of β-glucuronidase in tobacco leaf segments after sonication with the marker gene plasmid. Stable transfection was obtained with salmon sperm DNA as a carrier for sonication with the plasmid, but not in controls with plasmid or sonication alone. Wyber et al.[77] reported a 20-fold enhancement of transfection for 20 kHz sonication of yeast cells together with LEU2 reporter gene plasmids plus salmon sperm DNA (as a carrier DNA) relative to controls. Cavitation activity was measured by the iodine release assay, and correlated with loss of cell colony forming units (viability).

Since sonication treatment at 20 kHz is not commonly used in medicine (except in certain surgical procedures such as liposuction), the potential for DNA transfer with medically relevant MHz frequency ultrasound is of considerable interest. Kim et al.[7] developed an exposure system with 1 MHz ultrasound delivered into culture flasks and plates. Primary fibroblasts from rat hind limb muscles were cultured and exposed with a β-galactosidase control vector or with a plasmid (pMC1neo poly(A)) containing an antibiotic resistance gene. Variations in transfection efficiency were found with temperature, pressure amplitude above 0.2 MPa, exposure duration, and on–off timing. Highest transfection rates were found for about 0.4 MPa and 20 s exposures, with about 50% survival. Stable transfection rates averaged 0.34% of surviving cells. Treatment at 3.5 MHz resulted in no DNA transfer, which implicates cavitation-induced sonoporation as the mechanism. The involvement of cavitation was further demonstrated by augmenting nucleation with Albunex® ultrasound contrast agent.[10] A human condrocyte cell line was transfected with a marker plasmid coding for the green fluorescence protein (GFP) using 1 MHz ultrasound transmitted into culture dishes. Scoring of fluorescent cells was accomplished after cell harvest by flow cytometry. Transfection increased above an apparent threshold of 0.12 MPa (spatially averaged over the culture dish) and reached about 50% of viable cells at 0.41 MPa for 20 s (Fig. 9.2).

Seeding of the medium with 10% gas-body-based ultrasound contrast agent was required to promote transfection of CHO cells at low amplitudes in the rotating tube exposure system, clearly demonstrating that the transfection was mediated by ultrasonic cavitation.[8] Luciferase reporter plasmid at 20 μg/mL was added to the suspension during exposure, and cells were assayed for proliferation ability and for luciferase gene expression 2 days after exposure. Cell proliferation was greatly reduced above the cavitation threshold. Luciferase production was significant for 0.20 MPa exposure, and reached 0.33 ng per 10^6 cells at 0.8 MPa exposure. The luciferase production was greater for

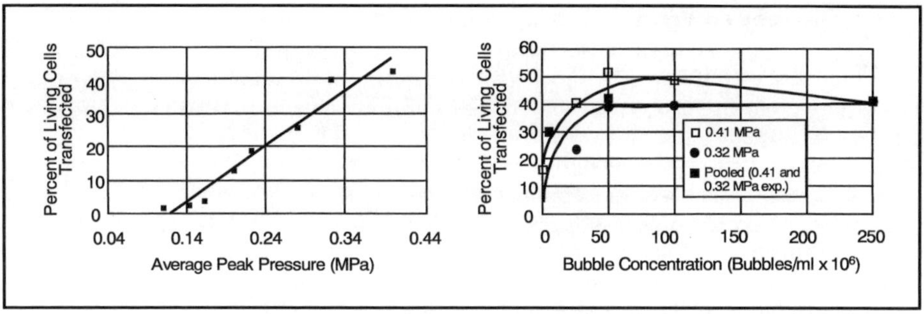

Fig. 9.2. Transfection rate of living cells after ultrasound exposure as a function of average peak pressure of the 20-s burst of 1 MHz ultrasound (left panel). Transfection rate of living cells after exposure plotted against Albunex concentration at the time of exposure (right panel). Reproduced with permission from Greenleaf et al.[10]

cells exposed in medium supplemented with serum than for cells exposed in serum-free medium.

Tata et al.[9] examined GFP marker plasmid transfer in two human cancer cell lines LnCap and PC-3. Ultrasound treatment was similar to that used in physical therapy with 932 kHz frequency and a 20% duty cycle. The ultrasound beam was directed upward into culture dishes with an average intensity of 0.33 W cm^{-2}. Loss of cell viability was greatest for continuous or low pulse repetition frequency (PRF) exposure, but remained near control levels for PRFs above 500 Hz. Similarly, as indicated by fluorescent- labeled plasmids in viable cells, the uptake of the DNA was greatest from continuous to 200 Hz PRF. Gene expression followed a similar pattern, with maximal expression for 50–150 Hz PRF, and no increase in GFP expression for a clinically relevant PRF of 1 kHz. This response pattern likely reflects the variation in occurrence and amount of cavitation activity in this system.

Lithotripter shock waves are also effective for in vitro transfection of cell suspensions. Expression of β-galactosidase gene in a reporter plasmid was observed in 0.1–0.5% of viable plated cells after 48 h.[78] The expression increased with increasing concentration of DNA during exposure and with increasing numbers of shock waves. The effect was suppressed by application of 10 MPa overpressure, which is a positive test for cavitation. The method was applicable to many different cultured cell lines including HeLa cells, mouse L-M fibroblasts, monkey kidney cells, and L1210 mouse leukemia cells. The GFP reporter plasmid was utilized by Miller et al.[79] to show the numbers of cells transfected by shock waves in vitro. A plasmid solution was added to the B16 melanoma cells in suspension at 20 μL mL^{-1} and 0.2 mL of air was retained in the 1.2-mL exposure chambers. After 800 SW in vitro, cell counts after 2 days of culture were only 5.3% of shams, with 1.4% of the cells expressing GFP. An important feature of the in vitro results was that the trends for transfection appeared to follow the same favorable trends noted previously for fluorescent Dextran loading of erythrocytes.[64]

The transfection efficacy of lithotripter shock wave treatment was compared to 1.18 MHz focused ultrasound exposure by Huber et al.[80] While lithotripter shock waves induced up to an eightfold increase in marker gene expression in the remaining 5% viable cells (relative to controls), focused ultrasound induced up to an 80-fold increase at 45% viability. Viability decreased and DNA transfer increased for increasing pressure amplitude up to 5 MPa at 40% duty cycle, and increasing sonication time up to 10 min. For varying PRF, 100 Hz appeared to be most effective. These results suggest that focused ultrasound could compete with lithotripter shock wave treatment in ultrasonically aided gene transfection, although in vitro results may not be faithfully reflected for in vivo conditions. Huber and Pfisterer[81] examined this question using 1.18 MHz focused ultrasound on prostate tumor cells in vitro and in vivo (see below for in vivo results). In vitro, 1 MPa exposures with 4 ms burst duration and 100 Hz PRF produced 55- to 220-fold increases in β-galactosidase expression in different cell lines including human melanoma and Dunning rat prostate tumor cell lines.

Ultrasound might be useful for gene therapy in cardiology applications. Lawrie et al.[82] cultured porcine vascular smooth muscle cells and endothelial cells and performed luciferase marker plasmid transfer with naked or lipid-charged DNA with or without 1 MHz ultrasound exposure of the monolayers. The transfection period was 3 h in duration, while ultrasound was applied at 0.4 W cm^{-2} only for 60 s, 30 min into the transfection period. The ultrasound application enhanced subsequent luciferase expression by 7.5 times for naked DNA, 2.4 times for lipid-charged DNA in the muscle cells and 3.3 times for lipofection of endothelial cells. Substantial heating, briefly reaching 51°C, was produced by the ultrasound but simulation of this heating had no effect on the subsequent luciferase expression. However, the ultrasound, but not the heating alone, induced a reduction in cell proliferation, which implicates the cavitational mechanism in the results. This implication was confirmed in experiments with 10% added Albunex® or Optison™ ultrasound contrast agents.[13] Up to 300-fold enhancement over naked DNA transfection was obtained for 956 kHz ultrasound. High-pressure amplitude exposure (about 2 MPa) was further associated with hydrogen peroxide production, which is a test for inertial collapse of the bubbles with free-radical production. However, free-radical generation is not required for transfection, since the hydrogen peroxide production was eliminated for lower exposure at about 1 MPa, for which the transfection efficiency was retained. The addition of Optison™ also enhanced lipofection (Tfx-50 lipoplexes) and polyplexes (LT-1) of the smooth muscle cells.

Manome et al.[83] have examined transfer of naked DNA into tumor cells in vitro and in vivo using 1 MHz continuous ultrasound (see below for in vivo results). In vitro, 30-s ultrasound treatment enhanced β-galactosidase and antibiotic-resistance reporter gene expression in mouse adenocarcinoma cells both in transient and stable transfection assays. Stable transfection for 10^5 cells cultured for 10 days with geneticin selection averaged 34 cells for 30-s ultrasound exposure.

A clinical spectral Doppler ultrasound system, as well as a laboratory system, was used by Koch et al.[84] to enhance transfection by cationic liposomes. The GFP reporter plasmid was used with rodent and canine glioma cells. For the clinical system, exposure was performed at 2 MHz, 0.5 W cm^{-2} average intensity in 12 well culture plates, and fluorescent cell counts made 24 h later. Up to 32.7% transfection was obtained for 90-s exposure, compared to 7.4% in controls. For the laboratory system, cells were grown on polypropylene membranes and exposed in a chamber with an ultrasound absorber to simulate free-field exposure conditions. If exposures were performed at 0.6 MPa pressure amplitude which did not produce cavitation (indicated by a lack of subharmonic emission), then no effects were found on viability or transfection. However, if Levovist® (Schering AG, Berlin) ultrasound contrast agent was added to the exposure chamber at 2–200 μg/mL, significantly increased transfection was observed. In addition, cell viability was reduced, for example, to 39.3% for the higher contrast agent concentration. Although the contrast agent was not used with the spectral Doppler system, it seems likely that cavitation activity was responsible for the observed effects.

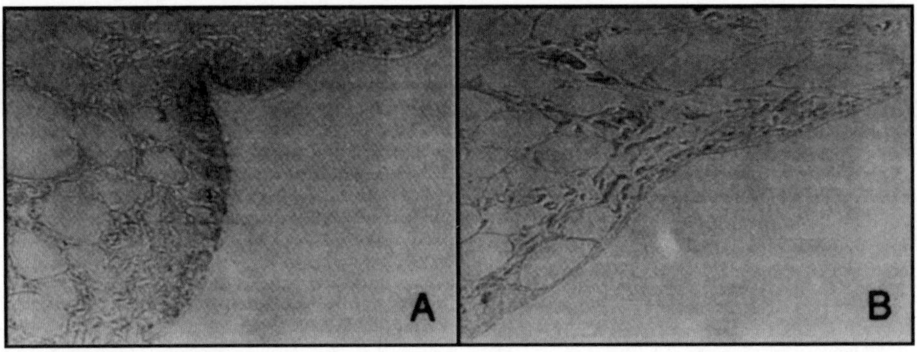

Fig. 9.3. Photomicrograph of rat knee joints after injection of a β-galactosidase plasmid and exposure to ultrasound in the presence of the contrast agent Albunex. Three out of four exposed knees showed β-galactosidase expression (A), while no expression was seen in control, unexposed (B) joints. Reproduced with permission from Kim et al.[7]

DNA Delivery in Vivo

The observation of transient permeabilization by shock waves leads to a suggestion by Delius et al.[85] that extracorporeal shock wave lithotripsy might be useful for gene therapy. On the basis of the in vitro results at higher frequencies, and even diagnostic ultrasound with contrast agents, many forms of ultrasound appear to be capable of DNA transfer to varying degrees depending on the cavitation activity obtained. However, it is well known that cavitation activity is minimized in vivo relative to in vitro conditions. Therefore, actual in vivo testing is required to accurately assess the potential of ultrasound-enhanced gene therapy.

Kim et al.[7] tested the possible extension of their in vitro experiments to in vivo conditions. The β-galactosidase reporter plasmid was injected into both knee joints of rats, and one joint was treated with 1 MHz plus 30 kHz ultrasound at 0.4 MPa and 40 kPa, respectively, for 1 min. Three of four treated knees showed reporter gene expression after 4 days, while none was detected in the unexposed joints, as shown in Fig. 9.3.

In vivo testing of the potential for enhanced cancer gene therapy by lithotripter shock wave exposure was performed using the luciferase marker plasmid in the B16 mouse melanoma tumor by Bao et al.[86] Luciferase reporter vector was injected with a concentration of 2 mg mL^{-1} at 10% of the tumor volume. Air at 10% of tumor volume was also injected after the DNA in some tumors to enhance acoustic cavitation activity. Shock waves were applied at 5.2 MPa peak negative pressure amplitude to the tumors of mice placed in a waterbath to assure free-field conditions. Exposure to 800 shock waves, followed by immediate isolation and culture of tumor cells for 1 day, yielded 1.1 (0.43 SE) pg/10^6 cells for plasmid injection only and 7.5 (2.5 SE) pg/10^6 cells for plasmid plus air injection. Significantly increased luciferase production, relative to shams, occurred for 200, 400, 800, and 1200 shock wave treatments with plasmid and air injection. Exposure with the isolation of tumor cells delayed for a day to allow gene expression within the growing tumors gave increased luciferase production for 100 and 400 shock wave exposure without and with air injection. For immediate harvest, expression was enhanced roughly 15-fold relative to direct injection alone, and air injection gave a further sevenfold increase. For delayed isolation, a 350-fold enhancement was found for 400 SW plus air compared to

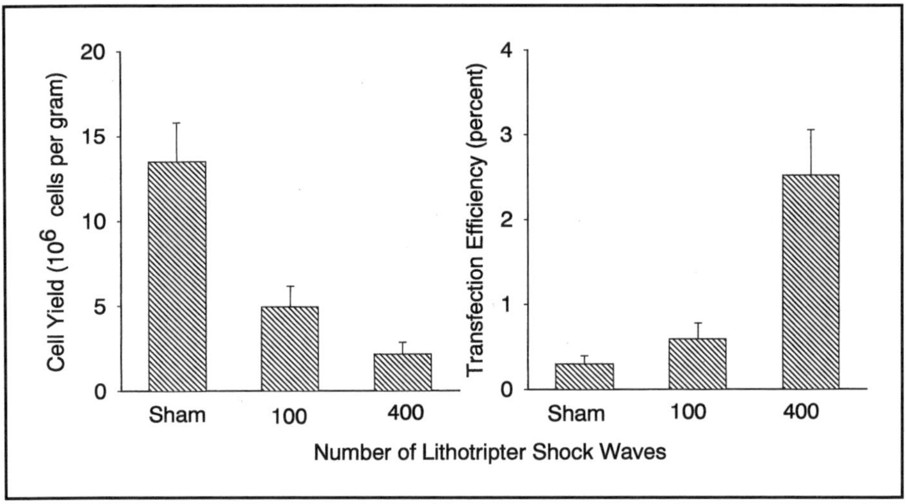

Fig. 9.4. Results for GFP transfection by lithotripter shock waves plus injected air in B16 mouse melanoma tumors. Cell yield after 2 days decreased with increasing exposure, but the percentage of the recovered viable cells which were fluorescent increased. Reproduced with permission from Miller et al.[79]

simple direct injection. The GFP reporter plasmid was utilized to elucidate the numbers of B16 melanoma tumor cells undergoing transfection.[79] The plasmid solution was injected intratumorally at 0.2 mg mL^{-1} DNA per milliliter of tumor and air was injected at 10% of tumor volume in vivo. For 400 SW exposure, viable cell recovery from excised tumors was reduced to 4.2% of shams and cell transfection was enhanced by a factor of about 8, reaching 2.5% of viable cell counts ($P < 0.005$ in t test). These results indicate tumor ablation plus enhancement of transfection for ultrasound SW treatment, as shown in Fig. 9.4.

Huber and Pfisterer[81] evaluated 1 MHz focused ultrasound enhancement of transfection of the β-galactosidase reporter plasmid into Dunning prostate tumors implanted in rats. Rats were exposed in a water tank to 1 MPa burst mode ultrasound (4-ms duration, 100 Hz PRF) after injection of 10 μg of the plasmid DNA. Staining of sections of exposed tumor revealed areas of β-galactosidase positive (blue stained) cells. DNA injection plus ultrasound produced a 10-fold increase in positive cells compared to intratumoral DNA injection alone, and a 15-fold enhancement in β-galactosidase protein assayed by ELISA. No transfection of tumor cells was obtained for intravenous injection of 100 μg of plasmid DNA with or without ultrasound exposure.

Manome et al.[83] also followed up on in vitro results with in vivo testing in colon carcinoma tumors implanted in mice. The tumors were injected with 200 μg β-galactosidase reporter plasmid and exposed to 1 MHz ultrasound at 10–20 W cm^{-2} 10 min later. Relative β-galactosidase activity 48 h after treatment increased approximately threefold for ultrasound, and was higher at the higher intensity. Expression peaked for 30-s exposure and declined for longer exposures up to 2 min. The mechanism of the enhanced transfection was not examined, but the intensities employed were sufficient to induce some cavitation activity at the injection site.

Ultrasound treatment appears to enhance transfection of tumor cells by cationic lipids in vivo.[87] Without ultrasound, most transfection obtained with CAT reporter or interleukin-12 coding plasmids complexed to cationic liposomes occurred in the lungs. For ultrasound treatment of the tumor before or after intravenous injection of the plasmid complex resulting in 3- 270-fold enhancement of gene transfer to the tumor, depending on conditions. The tumor DNA delivery increased with increasing DNA dose, and increasing exposure duration. The effect was maximized for ultrasound treatment up to 1.5 W/cm^2 about 1 min after injection of the DNA complexes. By the immunostaining of histological sections taken 15 min after treatment, the DNA delivery was localized to the endothelium within the tumor. The ultrasound treatment was termed sonoporation; however, the mechanism for this in vivo effect is not completely clear, because no cavitation enhancement was employed.

Ultrasound of relatively low power (2 $W\,cm^{-2}$ at 1 MHz for 5 min) has been applied to subcutaneous mouse tumors for use in therapy.[88] The treatment enhanced the effectiveness of chemotherapeutic bleomycin treatment and also the delivery of DNA plasmids to the tumor tissue. This ultrasound treatment was less effective for transfer and expression of a luciferase marker plasmid than was an electroporation procedure. It should be noted that the ultrasound was applied without added cavitation nucleation agents, and that the electroporation system required insertion of 10-mm wide steel strips into the tissue on two sides of the tumors, which complicates the interpretation of this comparison.

In vivo cardiovascular applications of sonoporation were also recently reported. Several studies addressed the possibility of preventing vascular proliferation (and thus of restenosis after angioplasty). In a porcine model of coronary artery injury, intravenous administration of an antisense nucleotide to *c*-myc in conjunction with PESDA (perfluorocarbon-exposed sonicated dextrose albumin) was associated with an apparent reduction in the degree of proliferation after 4 weeks.[89] The results are difficult to explain, since the addition of ultrasound at 1 MHz for 30 or 180 s did not induce an antiproliferative effect. These findings could eventually be explained by additional injury due to cavitation in those vessels that were insonated. However, the study opens the interesting possibility of enhanced transfer of oligunucleotides via echographic contrast agents alone.

Two other recent reports suggest a role for sonoporation in the prevention of restenosis. In a rat model of carotid artery injury, Taniyama et al.[17] showed a decrease in neointima formation after delivery of the antiproliferative gene p53 in combination with 25% Optison and insonation at 1 MHz for 2 min. Two weeks after delivery, treated vessels showed significantly lower intima to media ratio in comparison with both a negative control and p53 plasmid alone. Yamasaki et al.[15] have shown also in a rat carotid artery injury model reduced proliferation after administration of a decoy for the transcription factor E2F in conjunction with Optison (25%) and insonation at 1 MHz for 2 min. It should be emphasized, however, that these are preliminary results, and further studies are needed to confirm the findings.

A totally different approach to insonation was taken by Amabile et al.[18] In a rabbit model of femoral artery injury, plasmid-mediated transfection of blue fluorescent protein was enhanced by intravascular ultrasound at 2 MHz. Interestingly enough, the combination of plasmid and ultrasound was associated with higher expression of the reporter gene when compared even to adenoviral-based transfection.

Gene delivery to the myocardium of rats was obtained with harmonic mode diagnostic ultrasound, a microbubble contrast agent and a viral β-galactosidase vector.[14] The contrast agent was prepared in the laboratory and processed with the vector to attach the virus particles to the microbubbles. For exposure, 2 mL of this preparation was infused over a 30-min period via the jugular vein. Three frames from a 1.3 MHz transducer destroyed the microbubbles evident in the second-harmonic image, and three frame bursts were triggered to allow refill of the tissue between scans. After 4 days, expression of the reporter gene was assayed in histological sections and by measurement of enzyme activity. Staining and enzyme activity was detected in the myocardium after echocardiographic destruction of the microbubbles mixed with the viral vector at about 10 times the levels found in controls (bubbles plus ultrasound, no vector; bubbles plus vector, no ultrasound; vector alone, no bubbles, no ultrasound). β-Galactosidase activity was found in the livers of all animals receiving the viral vector. Cavitation activity was clearly responsible for the effect because the procedure involved destruction of contrast agent microbubbles; however, it is uncertain whether the viral vector was delivered by mechanical sonoporation or by some other process. Echocardiographic microbubble destruction followed by vector infusion produced about twice the gene expression of controls, suggesting that disruption of the endothelial barrier during microbubble destruction might be a factor in the enhanced viral transduction.

Conclusions and Prospects

Ultrasound causes perturbation of cell membranes through the interaction of cavitation bubbles with the cells. This can lead to lysis or to sonoporation, in which permeabilized cell membranes heal and allow the cell to survive. Sonoporation appears to be injurious to the cells, but affords an opportunity to deliver large molecules from the extracellular medium to the cytoplasm. Sonoporation is conveniently studied using large fluorescent Dextran molecules, with DNA delivery a special case. Although cavitation activity generally decreases for increasing ultrasound frequency, sonoporation has been demonstrated for 20 kHz ultrasonic probes, for lithotripter shock waves and even for megahertz frequency ultrasound as used for medical therapy. With augmentation of cavitation activity by adding gas-body-based contrast agents to the exposure medium, sonoporation and DNA delivery has been produced by diagnostic ultrasound in cell suspensions and in monolayer culture systems. In vivo situations typically show minimal cavitation activity due to a dearth of cavitation nucleation sites; however, by augmenting nucleation with the inclusion of injected air or microbubble contrast agents, sonoporation and DNA delivery has been shown in several tumor models and in normal tissues.

A clear need in gene therapy is for novel vector delivery methods. Several gene delivery methods are now available, including direct plasmid injection into tumors, DNA-lipid conjugates, and viral vectors, but none appears to be fully developed nor does any one method have universal applicability. Viruses yield high transfer efficiency, but are immunogenic, or even mutagenic, and infect many nontumor cells.[90,91] The efficiency found for DNA lipid conjugates in vitro does not necessarily improve efficiency in vivo over simple injection of naked DNA.[92] Direct plasmid injection is appealing for its simplicity and avoidance of viruses, but has yielded a low level of transfection in tumors.[86,93] Other novel methods include "gene gun" delivery of DNA on particulate projectiles,[94] and in vivo electroporation with implanted electrodes.[95]

This review has detailed the initial progress by several researchers toward development of ultrasound-enhanced gene therapy. Studies have shown ultrasound enhancement of DNA delivery over direct injection, lipofection, and even over viral transfection. Theoretically, sonoporation could help deliver genes with high spatial and temporal specificity. The importance of selective targeting in gene therapy cannot be overemphasized, considering the risks associated with systemic delivery of genetic vectors at high doses. Indeed, an effect that is desirable in one region (e.g., induction of apoptosis in tumor cells) is potentially fatal if generalized to other organs.

The development of "smart" bubbles that can recognize and attach to specific targets might enhance the spatial resolution even further, providing a real "ZIP code based" delivery of genetic material. In addition, ultrasound-induced bioeffects beyond sonoporation could enhance the effects of gene therapy within a single treatment modality. Delivery of DNA outside blood vessels by capillary rupture might be valuable to target cells of a specific tissue. In cancer gene therapy, the concomitant tumor ablation with lithotripter exposure might provide augmentation of a gene mediated antitumoral therapy. Further research should elucidate the potential of the ultrasound enhanced DNA delivery and identify the most promising therapeutic applications.

Acknowledgment

This work was supported in part by grant number CA83904 (D.L. Miller) from the National Institutes of Health.

Literature Cited

1. Nabel EG. Circulation 1995; 91:541-548.
2. Schneider MD, French B. Circulation 1993; 88:1937-1942.
3. Gershon H, Ghirlando R, Gutman SB et al. Biochemistry 1993; 32:7143-7151.
4. Chang DC, Reese TS. Biophys J 1990; 58:1-12.
5. Daniell H. Methods Enzymol 1993; 217:536-557.
6. Newman CM, Lawrie A, Brisken AF et al. Echocardiography 2001; 18:339-347.
7. Kim HJ, Greenleaf JF, Kinnick RR et al. Hum Gene Ther 1996; 7:1339-1346.
8. Bao S, Thrall BD, Miller DL. Ultrasound Med Biol 1997; 23:953-959.
9. Tata DB, Dunn F, Tindall DJ. Biochem Biophys Res Comm 1997; 234:64-67.
10. Greenleaf WJ, Bolander ME, Sarkar G et al. Ultrasound Med Biol 1998; 24:587-595.
11. Mukherjee D, Wong J, Griffin B et al. J Am Coll Cardiol 2000; 35:1678-1686.
12. Porter TR, Iversen PL, Li S et al. J Ultrasound Med 1996; 15:577-584.
13. Lawrie A, Briskin AF, Francis SE et al. Gene Ther 2000; 7:2023-2027.
14. Shohet RV, Chen S, Zhou Y et al. Circulation 2000; 101:2554-2556.
15. Yamasaki K, Tachibana T, Taniyama Y. Circulation 2000; 102:II-247.
16. Komamura K, Tatsumi R, Miyatake K. Circulation 2000; 102:II-561.
17. Taniyama Y, Tachibana K, Yamamoto K. Circulation 2000; 102:II-164.
18. Amabile PG, Waugh JM, Lewis TN et al. Circulation 2000; 102:II-147.
19. Fu YK, Kaufman GE, Miller MW et al. Radiat Res 1979; 80:575-580.
20. Mortimer AJ, Dyson M. Ultrasound Med Biol 1988; 14:499-506.
21. Dinno MA, Crum LA, Wu J. Ultrasound Med Biol 1989; 15:461-470.
22. Mihran RT, Barnes FS, Wachtel H. Ultrasound Med Biol 1990; 16:297-309.
23. Pohl EE, Rosenfeld EH, Pohl P et al. Ultrasound Med Biol 1995; 21:711-719.
24. Kaufman GE, Miller MW. Ultrasound Med Biol 1978 4:139-144.
25. Miller DL, Thomas RM, Buschbom RL. Ultrasound Med Biol 1995; 21:841-848.
26. Liebeskind D, Bases R, Mendez F et al. Science 1979; 205:1273-1275.
27. Miller MW, Miller DL, Brayman AA. Ultrasound Med Biol 1996; 22:1131-1154.
28. Kaufman GE. Ultrasound Med Biol 1985; 11:497-501.
29. Dalecki D, Keller BB, Raeman CH et al. Ultrasound Med Biol 1993; 19:385-390.
30. MacRobbie AG, Raeman CH, Child SZ et al. Ultrasound Med Biol 1997; 23:761-765.
31. Skyba DM, Price RJ, Linka AZ et al. Circulation 1998; 98:290-293.
32. Miller DL, Quddus J. Proc Natl Acad Sci USA 2000; 97:10179-10184.
33. Hynynen K, Chung AH, Colucci V et al. Ultrasound Med Biol 1996; 22:193-201.
34. Child SZ, Hartman CL, Schery LA et al. Ultrasound Med Biol 1990; 16:817-825.
35. Tarantal AF, Canfield DR. Ultrasound Med Biol 1994; 20:65-72.
36. Zachary JF, O'Brien WD, Jr. Vet Pathol 1995; 32:43-54.

37. Miller DL, Thomas RM. Ultrasound Med Biol 1995; 21:249-257.
38. Dalecki D, Child SZ, Raeman CH et al. J Acoust Soc Am 1995; 97:3165-3170.
39. Ramnarine KV, Nassiri DK, McCarthy A et al. Ultrasound Med Biol 1998; 24:575-585.
40. Tsirulnikov EM, Vartanyan IA, Gersuni GV et al. Ultrasound Med Biol 1998; 14:277-285.
41. Nilsson AM, Odselius R, Roijer A et al. Ultrasound Med Biol 1995; 21:833-840.
42. Tachibana K, Tachibana S. Circulation 1995; 92:1148.
43. Harrison GH, Balcer-Kubiczek EK, Eddy HA. Int J Rad Biol 1991; 59:1453-1466.
44. Saad AH, Hahn GM. Ultrasound Med Biol 1992; 18:715-723.
45. Price RJ, Skyba DM, Kaul S et al. Circulation 1998; 98:1264-1267.
46. Unger EC, McCreery TP, Sweitzer RH et al. Invest Radiol 1998; 33:886-892.
47. ter Haar G. Ultrasound Med Biol 1995; 21:1089-1100.
48. Bender LF, Janes JM, Herrick JF. Arch Phys Med Rehabil 1954; 35:555.
49. Duarte LR. Arch Orthop Trauma Surg 1983; 101:153-159.
50. Nyborg WL, Carson PL, Carstensen EL et al. Exposure Criteria for Medical Diagnostic Ultrasound: II Criteria Based on all Known Mechanisms. Report No. 140, National Council on Radiation Protection and Measurements, Bethesda, MD, 2001.
51. Nyborg WL, Carson PL, Miller MW et al. Exposure Criteria for Medical Diagnostic Ultrasound: I. Criteria Based on Thermal Mechanisms, National Council on Radiation Protection Report No 113, Bethesda, MD, 1992.
52. American Institute of Ultrasound in Medicine. J Ultrasound Med 2000; 19:69-168.
53. ter Haar G, Duck FA. (eds.). The Safe Use of Ultrasound in Medical Diagnosis, British Medical Ultrasound Society/The British Institute of Radiology, London, 2000.
54. Whittingham TA. In Ultrasound in Medicine, FA Duck, AC Baker, HC Staritt (eds.), Institute of Physics Publishing, Philadelphia, 1998, pp. 129–148.
55. Dalecki D, Child SZ, Raeman CH et al. Ultrasound Med Biol 1997; 23:777-781.
56. Roy RA, Atchley AA, Crum LA et al. J Acoust Soc Am 1985; 78:1799-1805.
57. Clarke PR, Hill CR. J Acoust Soc Am 1970; 47:649-653.
58. Fechheimer M, Denny D, Murphy RF et al. Eur J Cell Biol 1986; 40:242-247.
59. Fechheimer M, Boylan JF, Parker S et al. Proc Natl Acad Sci USA 1987; 84:8463-8467.
60. Johannes C, Obe G. Mutat Res 1997; 374:245-251.
61. Delius M. Shock Waves 1994; 4:55-72.
62. Gambihler S, Delius M, Ellwart JW. J Memb Biol 1994; 141:267-275.
63. Delius M, Adams G. Cancer Res 1999; 59:5227-5232.
64. Miller DL, Williams AR, Morris JE et al. Ultrasonics 1998; 36:947-952.
65. Zhong P, Haifan L, Xi X et al. J Acoust Soc Am 1999; 105:1997-2009.
66. Doukas AG, Flotte TJ. Ultrasound Med Biol 1996; 22:151-164.
67. Lee S, McAuliffe DJ, Zhang H et al. Ultrasound Med Biol 1997; 23:1089-1094.
68. Mulholland SE, Lee S, McAuliffe DJ et al. Pharm Res 1999; 16:514-518.
69. Miller DL, Bao S, Morris JE. Ultrasound Med Biol 1999; 25:143-149.
70. Ward M, Wu J, Chiu J. J Acoust Soc Am 1999; 105:2951-2957.
71. Ward M, Wu J, Chiu J. Ultrasound Med Biol 2000; 26:1169-1175.
72. Miller DL, Quddus J. Ultrasound Med Biol 2000; 26:661-667.
73. Joersbo M, Brunstedt J. Plant Cell Rep 1990; 9:207-210.
74. Joersbo M, Brunstedt J. Physiol Plantar 1992; 85:230-234.
75. Joersbo M, Brunstedt J. J Virol Methods 1990; 29:63-69.
76. Zhang L, Cheng L, Xu N, Zhao N et al. Biotechnology 1991; 9:996-997.
77. Wyber JA, Andrews J, D'Emanuele A. Pharma Res 1997; 14:750-756.
78. Lauer U, Burgelt E, Squire Z et al. Gene Ther 1997; 4:710-715.
79. Miller DL, Bao S, Gies RA et al. Ultrasound Med Biol 1999; 25:1425-1430.
80. Huber PE, Jenne J, Debus J et al. Ultrasound Med Biol 1999; 25:1451-1457.
81. Huber PE, Pfisterer P. Gene Ther 2000; 7:1516-1525.
82. Lawrie A, Briskin AF, Francis SE et al. Circulation 1999; 99:2617-2620.
83. Manome Y, Nakamura M, Ohno T et al. Hum Gene Ther 2000; 11:1521-1528.
84. Koch S, Pohl P, Cobet U et al. Ultrasound Med Biol 2000; 26:897-903.

85. Delius M, Hofschneider P, Lauer U et al. Lancet 1995; 345:1377.
86. Bao S, Thrall BD, Gies RA et al. Cancer Res 1998; 58:219-221.
87. Anwer K, Kao G, Proctor B, Anscombe I et al. Gene Ther 2000; 7:1833-1839.
88. Tomizawa M, Ebara M, Saisho H et al. Cancer Lett 2001; 173:31-38.
89. Porter TR, Kricsfeld D, Xie F. J Am Soc Echocardiogr 2000; 13:434.
90. Cooper MJ. Semin Oncol 1996; 23:172-187.
91. Berinstein NL. Biological therapy of cancer, In The Basic Science of Oncology, 3rd edn., IF Tannock, RP Hill (eds.), McGraw Hill, New York, Chap. 18, 1998.
92. Nomura T, Nakajima S, Kawabata K et al. Cancer Res 1997; 57:2681-2686.
93. Yang JP, Huang L. Gene Ther 1996; 3:542-548.
94. Rakhmilevich AL, Turner J, Ford MJ et al. Proc Natl Acad Sci USA 1996; 93:6291-6296.
95. Nishi T, Yoshizato K, Yamashrio S et al. Cancer Res 1996; 56:1050-1055.

Index

E

F

G

H

I

L

M

N

O

P

Q

R

S

T